The
Plumber's Bible

The Plumber's Bible

A HomeOwner's Bible

SCOTT WILSON

DOUBLEDAY & COMPANY, INC., GARDEN CITY, NEW YORK
1981

Library of Congress Cataloging in Publication Data

Wilson, Scott.
  The plumber's bible.

  1. Plumbing—Amateurs' manuals.   I. Title.
TH6124.W54   644'.6
ISBN: 0-385-11211-4
Library of Congress Catalog Card Number 76–42420

Contents

The
Plumber's Bible

CHAPTER ONE

How It Works

All plumbing systems are alike in that the reason water flows out of a faucet when you open it is that pressure is applied by one means or another to the water in the pipe.

If your home is connected to a municipal water main, the pressure exerted on the water in your pipes is produced by a pump and sometimes by a water tower too, both of which may be dozens of miles away. Singly or in tandem, the pump and the tower provide a fairly constant pressure on the water at their end of the main, which is merely a large pipe (actually a system of interconnected pipes) that distributes water to the buildings in the community. Each house so served is connected to the main by a smaller-diameter pipe called a water service line.

The water service line or pipe runs underground from the city water main, which is generally a distance below the center of the road in front of the house, through a curb stop (valve), which is also underground and then to a large valve called the main valve or house valve just inside the basement or cellar. From here the cold water is led by pipe through a water meter (if there is one) and then on to the various fixtures (sinks, tubs, etc.), appliances (washing machine, etc.), furnace and sill cock (water connection for the garden hose). There may or may not be additional valves in the pipes leading to groups of fixtures and individual fixtures.

If you have your own well, the pressure necessary to drive the well water through the pipes and up and around the house will come from a pump. To reduce the wear and expense of operating the pump continuously, any of several strategies are

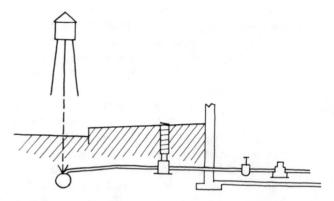

1. Water under pressure is supplied to the water main by a system of water towers and/or pumps. Water from the main flows into the house through the water service pipe, the curb valve, the main valve and the water meter. Shutoff can be accomplished at the curb valve and the main valve.

used. One is the use of a water tank atop the attic floor or a tower. A float-operated switch in the tank controls the pump, turning it on when the tank is near empty and stopping it when the tank is full. The height of the tank provides the almost constant pressure by virtue of the water's weight. Another system involves a tank closed at one end. The pump drives water into the tank against a volume of air. The air is compressed and the motor is automatically shut off when the tank is about three-fourths filled with water. When a faucet is opened the compressed air expands and forces the water out. When the tank is nearly empty the motor automatically kicks in. In newer designs, a rubber diaphragm divides the tank into two sections—air and water—and prevents the air from dissolving into the water, which would necessitate the tank being drained every few months or so.

STOPPING THE FLOW OF WATER

To stop the flow of hot or cold water out of a defective faucet or a break in a pipe, two steps are necessary: the pressure on the water must be removed and the water in the affected section of pipe must be drained.

To remove the pressure, trace the pipe back from the leaking faucet or pipe break to the first valve. You may find this directly beneath the sink or lavatory (where sensible design calls for it) or farther along the pipe. If you fail to find a valve in either place, or in the event of an emergency, the main valve—the valve next to the water meter—can be closed. Its handle is turned clockwise (to the right as you look down at it) until it cannot be turned anymore. A gate valve should be installed here; the gate valve (discussed in the following chapter) may require a dozen turns for closure.

Incidentally, it is advisable to locate this valve *before* you suffer an emergency, and to close it experimentally to make certain that you can close it by hand and that it has not "frozen" in position through disuse. If it is stuck, use a wrench to open and close it a few times. This should loosen it up. Check its operation by observing the flow of water out of an open faucet, after draining the line (to be discussed shortly). Should you be unable to stop the flow of water by reasonably tightening the main valve. it is defective and needs attention. (See Chapter Eight.)

Curb-stop closure—Should the main valve leak, the curb stop (valve) must be closed. This can only be done by means of a long-handled socket wrench owned and operated by the city waterworks. A phone call will fetch them. To facilitate their work it is wise to find the curb-stop cover. This is a hinged metal cover that hides and protects the vertical pipe above the curb stop. Since the curb stop is seldom used, its cover is often overgrown with grass.

Removing local water pressure—Should your home be supplied by a local well system, water pressure is removed by shutting off the pump, closing the main valve leading to the gravity tank (if there is one) or releasing the pressure in the air tank. The latter is done by simply draining the system.

Draining the system—At this juncture, the pressure on the water in the pipes (hot and cold) has been removed and no more water can enter the pipes because the valve is closed. However, there is water still present in the pipes (and pressure tank, if there is one). Should you open a faucet, water will run out. If there is a leak in a pipe, water will continue to emerge, though considerably more slowly. Therefore, the system has to be drained as well as the main valve closed before work is attempted.

Draining is accomplished by opening one cold- and one hot-water faucet in the upper reaches of the building, following which one or more hot and cold faucets are opened in the basement. When all water ceases to flow you know there is

2. You will find the main valve between the water meter and the foundation wall.

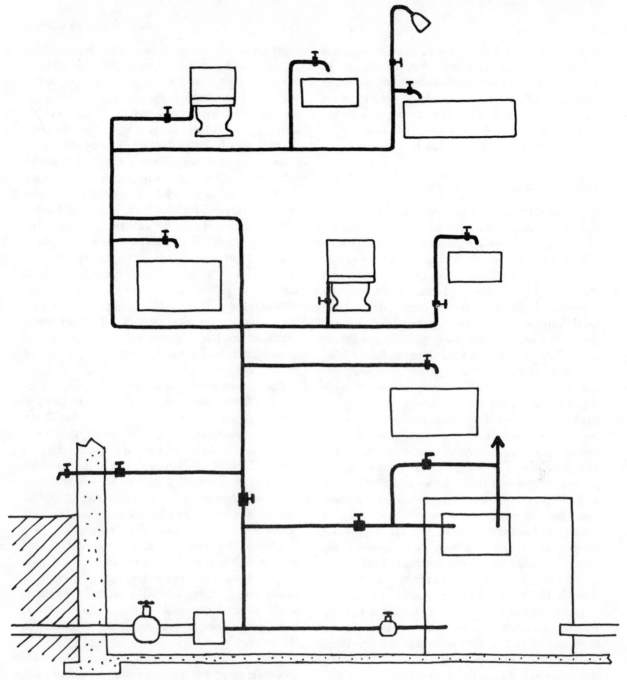

3. Basic cold-water system. Pipes are shown as short and straight, but obviously they can run in whatever direction is most convenient.

no more water in the pipes and that the main valve is holding properly.

There are two precautions and one exception to the foregoing. Do not open the small faucet you may find near the bottom of your furnace. Doing so will drain the boiler and there is no need to do that. Do not open the faucet at the bottom of a hot-water heating tank. Doing so will drain the tank, and if you open this faucet without first opening an upstairs hot-water faucet, air pressure may collapse your tank.

The exception is simply this. If you close the

valve directly beneath a faucet, there is no need to close the main valve and drain the system. If you close a valve intermediate between the main valve and the faucet there is still no need to drain the lines. It doesn't matter if water remains in the pipe or pipes, so long as the water is below your leak or the faucet you are going to work on.

THE SOURCE OF HOT WATER

No matter what type of device may be used in your home to produce hot water, this device is always connected to the cold water line, which means that the pressure driving the hot water out of your faucets is the same city pressure, and water, supplied to the cold-water pipes. Or, the pressure on the hot water is derived from your local well pump. Therefore, when you shut off the main valve or local pump you are also shutting off the flow of cold water into the heating device and, naturally, the flow of hot water out.

Hot water temperature and quantity—Assuming your hot-water heating device is in proper working condition, the temperature of the water coming out will depend on the setting of the thermostat, the temperature of the incoming water and the quantity of water you draw over a given period of time. All things being equal, it is normal for your hot water to be hotter in the summer than the winter because incoming water is warmer in the summer. Also, you can expect the temperature of your hot water to drop when you exceed the capacity of your hot-water heater; this is not difficult to do, as the capacity of the hot-water heater in most homes is limited.

There are two types of hot-water heaters. They will be described here to enable you to get the most out of each type and to prevent you from wasting time and money seeking more hot water than your unit is capable of providing.

One type is called "tank" because it consists of a large tank holding 35 to 100 gallons of water. It may be heated by electricity or gas. In the first case there will be two pipes emerging from its top. One will be the cold-water pipe feeding cold water into the tank. The other will be the pipe carrying hot water out. Electrical tanks may be positioned anywhere: in closets, under sinks and so on. The gas-fired tank will have the same two

pipes on top plus a third pipe along the side supplying the necessary heating gas. In addition there will be a vent pipe on top. This will run to the open air to remove gas fumes. Gas units must be separated by several feet from the nearest combustible surface or material. The flue pipe must always be in place and nothing should be placed near or on top. There is always a flame inside the gas tank; accordingly, proper precautions must be taken when you work on or near one. In terms of operation there is little difference between tanks heated by electricity or gas. The foregoing was given as a safety measure.

Tank heaters carry two ratings: capacity and recovery time. The first rating is the tank's total capacity in gallons and is somewhat deceptive: although the tank may be completely filled with hot water, you cannot use it all. Cold water enters as the hot is drawn off. Thus, the useful capacity of a tank is no more than 70 per cent of its total capacity.

The second rating, recovery time, means literally that: how long it takes the tank to come back up to preset temperature after it has been drained. Time varies with the size of the tank's heating unit and the temperature of the incoming water. Typically a Rheem 50-gallon, gas-fired tank will produce 37.8 gallons of hot water in one hour when the water's temperature has to be raised 100 degrees, and 70 gallons when the temperature difference is only 60 degrees.

While a 50-gallon tank may sound like a lot of tank, it isn't when you stop to consider just how much hot water we use today. On an average, a five-minute shower bath will consume 15 gallons of hot water. A dishwasher takes about 4 gallons. A ten-pound load of laundry needs 40 gallons of hot water.

Thus, if your hot-water supply is suddenly inadequate, and your home is equipped with a tank heater, it may well be that your family's washing habits have changed. Although your equipment may have sufficed when the hot water was drawn sequentially with time for the unit to recover, now that everyone is showering or bathing in the morning, there isn't enough.

All tank heaters become internally coated with lime with the passage of time. As the lime layer increases in thickness recovery time also in-

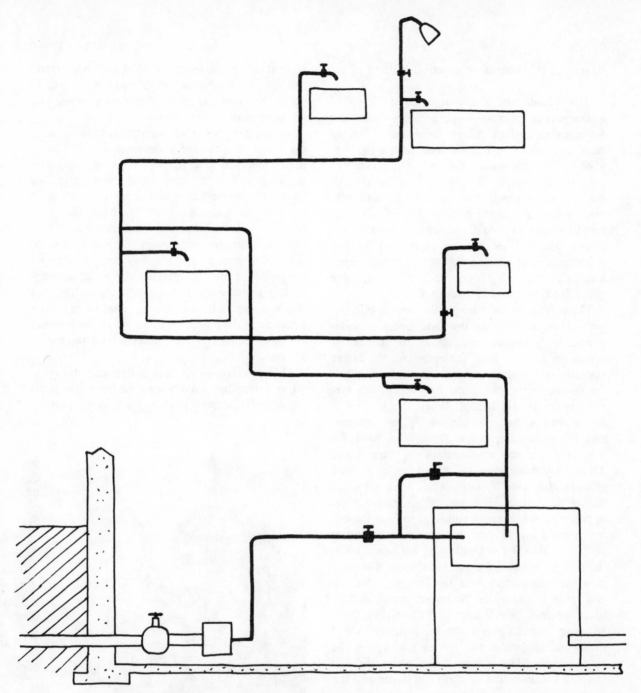

4. Basic hot-water system. Cold water enters the heating device and emerges hot. Hot-water pressure derives from the pressure on the cold water.

creases. The tank holds as much and the water gets just as hot, but it takes the tank longer and longer with the passage of years to come up to temperature. To test, open a hot water faucet and let the water flow until it runs cold. Then draw no hot water whatsoever and measure just how long it takes the tank to come up to temperature.

Pushing the thermostat setting up helps a little, but not much. When the tank has slowed to where you and your family can no longer wait for it, replace it. There is no practical method of repairing it.

The second type of hot-water heater is called instantaneous because it purportedly heats the water in an instant. There is no tank. Instead there is a coiled tube positioned within the furnace boiler. One end of the coil is connected to the cold-water pipe. The other end is the start of the hot-water piping system. (If you carefully place your hand on the ends of this coil you will find that one end is cold, the other is hot.) In addition, you will see a valve connected by two short lengths of pipe to the hot and cold pipes leading to the coil. This valve is a mixing valve and will be described in this chapter.

The coil heater differs from the tank in two important ways. Whereas the tank operates independently, going on and off as its thermostat senses the need for heat, the instantaneous heater operates in conjunction with the furnace. When the furnace is operating the coil within the furnace boiler is also heated. When the furnace is not operating, for example during the summer, and the water thermostat senses the need for heat, the furnace is automatically turned on. Thus the instantaneous hot water heater is more efficient and costs less to operate than the tank during the winter. However, in the summer the opposite is true, because you have to heat the entire furnace just to heat the immersed hot water coil. And, should you be drawing hot water when the furnace is off, there will be a noticeable drop in water temperature until the furnace starts up and heats the water in the boiler, which in turn heats the coil. This is normal. In addition, there will also be a drop in water temperature should the room thermostat call for heat and turn the circulator on. This drives boiler water through the cool radiators and so cools the boiler and the hot water coil. Again, this is normal.

However, once the furnace is running and the heater coil is hot, water emerging from a hot-water faucet will flow hot continuously. However, it will cool if you draw too much, which would be the case if all the hot-water faucets in the house were opened at one time.

In other words, with a tank system you can draw a tank's capacity of hot water at one gulp if you will. All of it will be hot until the tank is nearly empty. Then you have to wait for the tank

to heat more (recover). With a coil heater you can draw an unlimited quantity so long as you do not try to draw too much at one time. When you do, water temperature drops.

To prevent this from happening, many plumbers install a pressure reducer in the hot-water line. Usually it is a washer in the mixing valve. Sometimes it is a valve in the hot-water line itself. The presence of either or both of these obstructions in the hot-water line limits the quantity of hot water that can pass, no matter how many faucets are wide open. By doing this the hot water emerging from any and all faucets remains hot, but the quantity drops off as more faucets are opened. If the hot water coming out of your second-floor faucet disappears when someone in the basement turns the clothes washer on, this is what has been done to limit the use of hot water.

Mixing valve—This is a temperature-compensating valve connected across the hot-water heating coil. Its purpose is to mix a varying quantity

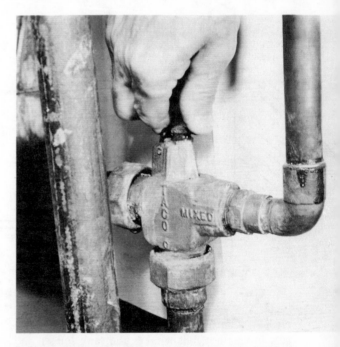

5. You will find the mixing valve close to the heating furnace. Turning it one way increases the temperature of the water entering the hot-water system. Turning the valve handle the other way decreases the temperature of the water.

of cold water with the hot so that the hot water is neither too hot nor cold. This is necessary because most boilers are operated around 200° F., and 145° F. is about as high as you want water to enter the hot-water pipes. A temperature of about 135° F. is the beginning of scalding, so if water enters the pipes in the basement at 145° F. it is just about as hot as you can safely stand when it emerges from an upstairs faucet. Also, in mixing you increase the quantity of hot water you can draw.

Liming—Instantaneous heaters are plagued to a greater degree by the formation of lime inside and outside the submerged coil than are tank heaters. This is because the rate of liming increases with increasing water temperature as well as with the quantity of minerals dissolved in the water, and instantaneous coils always run 40 or so degrees hotter than tanks.

As the coil limes up its ability to heat the passing water diminishes. Initial water temperature will be just as high, but when you open a second or third faucet its temperature will drop. As the coil limes you can compensate to some degree by adjusting the mixing valve so that it passes less and less cold water. However, when the handle is all the way to one side, there is no more compensation available and you are risking a serious burn every time you open a hot-water faucet. The first splash may be boiling hot. Open the faucet wide and the water runs lukewarm. At this point you may be advised to have your oil-burner serviceman turn the boiler thermostat up and so increase the boiler water temperature and at the same time the temperature of the immersed coil. Don't do it: this will increase the danger of a burn. The wiser course of action is to have the coil cleaned with a solution of hydrochloric acid and water, which sometimes works, or to replace the coil, which always works.

DRAINS

Drains are pipes that carry dirty water and muck *away* from sinks and toilets. Drainpipes differ from hot- and cold-water pipes in that drains are always larger in diameter than the largest hot- or cold-water pipe. Also, drains are always pitched downward: that is, if you start at a sink or a toilet and follow the drain, you will find that it is always going downhill. This is because waste, which is dirty sink water, and soil, which is toilet discharge, flow by gravity alone, whereas both the hot and the cold water flow by virtue of the pump pushing against the water at the far end of the city water main. Thus, if you were to make a small hole in a water pipe the water would squirt out. A similar hole in a drainpipe would merely cause a drip.

This is an important difference. In an emergency you can seal a hole in a drainpipe with a plug of wood or a wrapping of tape. A similar plug or wrapping won't hold a drop when used on a water pipe. The reason is the difference in pressure. The pressure on waste or soil in a drainpipe won't run over a dozen pounds per square inch at most. Water pressure usually ranges from 35 to 50 pounds, but can be as high as 125 pounds in some valley locations.

Sewer gas and traps—All the drainpipes in your home are connected to one large, main drainpipe called the house drain, which passes out through the foundation wall, whereupon it is called the sewer pipe. From here it continues on to either a septic tank on your property or the municipal sewer pipe somewhere out under the road.

Both the city sewer line and the septic tank are sources of sewer gas, which is a foul-smelling, complex, poisonous and explosive mixture produced in the main by the action of anaerobic bacteria, which live in sewer pipes and septic tanks. Some cities have long used this gas for heating and similar purposes. To prevent the sewer gas from entering your home, each drainpipe leading from a fixture (sink, tub, etc.) has been provided with a trap. This consists of a loop of pipe in which a slug of water remains after the main body of waste has passed through. The slug or plug of water prevents the low-pressure sewer gas from coming up and out the drain openings in your sinks and tubs and washing machines. The toilet does not have a trap in its drainpipe. Instead, a trap is built into each toilet.

In addition, most home drains are equipped with a large trap in the house drain just before it exits the building. This trap is often called a running trap. Thus, sewer gas is blocked from enter-

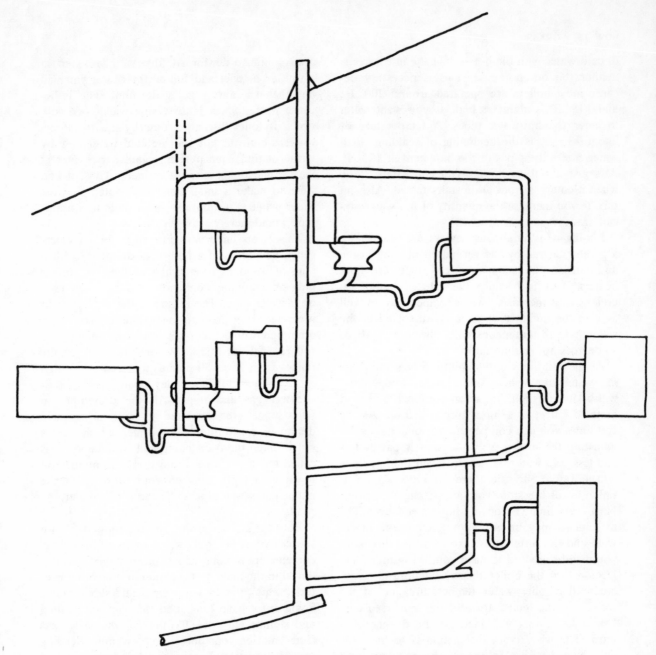

6. Basic drain system. Note that all of the fixtures below the top floor are vented by connection to the soil stack. Note that connection is made above the top-floor fixtures.

ing the building by two plugs of water, one in the main drain line and another in each of the connecting drain lines.

Vents—When a large quantity of water rushes down through a trap there is a tendency for the water in the trap to follow due to the siphon effect that occurs. To prevent this from happening, all traps except the running trap are vented. Venting is generally accomplished in one of two ways. A vent pipe may be connected to the drainpipe at a point just downstream of the trap. The vent pipe runs up through the roof and is always open to the air. Or the trap may be vented by a nearby drain stack, which is a large-diameter drainpipe stood on end. Its top is open to the air, its bottom leads to the house drain, through the running trap and on to the sewer. In some designs, depending on the way the pipes can be most conveniently positioned, the vent pipe does not run up through the roof, but is connected to the stack at a point some distance above the trap it vents.

In addition, many house drains are equipped with a fresh-air vent. This is a large-diameter pipe, one end of which is screened and projects through the foundation wall to the open air. The other end is connected upstream of the running trap. The fresh-air vent admits air to the lower end of the drain system. The soil heats the air, which then moves upward and out the upper, open ends of the vent pipes and stack. The constant passage of air through the drainpipes discourages the growth of the anaerobic bacteria that naturally line the drainpipe and so reduces the smell and the corrosive acids they would otherwise produce.

CHAPTER TWO

Plumbing System Components

PIPES

Several kinds of pipe are used in the average home plumbing system. If the building is more than thirty years old, most likely the hot and cold water are carried in galvanized wrought-iron pipe. Newer homes generally have copper tubing for the water lines, and an increasing percentage of recently constructed homes utilize plastic pipe for their water lines.

Drainpipes are usually of thin-wall brass directly following the sink and tub, and just as usually are connected to galvanized iron pipes just inside the wall. The galvanized pipes run down to the cellar or basement, where they are in turn connected to a larger-diameter cast-iron pipe. The cast-iron pipe will be carried through the basement and outside the building for a distance of about six feet, whereupon the cast-iron pipe will be connected to vitrified clay tile for its final run to the municipal sewer or local septic tank. In some towns less expensive and less desirable asbestos cement pipe (transite) will be used in place of the clay.

Newer homes use drain piping of copper inside the walls and down into the basement, where the copper is connected to the heavier, stronger and cheaper cast iron. In still newer homes only the short section of drainpipe immediately connected to the fixture drain is of metal; the rest of the run, even out to the municipal sewer, is of plastic pipe.

PIPES AND TUBING

A pipe is called a pipe when its wall is about ⅛-inch thick and a tube when its wall is less than $\frac{1}{16}$-inch thick. Thus, galvanized iron pipe is pipe because it has a "thick" wall. Similarly dimensioned copper is also called pipe, as is brass. But copper with a wall less than $\frac{1}{16}$ inch in thickness is called tubing, as is the stainless steel tubing recently introduced for home use.

PIPE SIZE

A pipe is nominally sized according to its inside diameter. Thus, a 1-inch galvanized pipe has an external diameter of close to 1⅜ inches and a 2-inch pipe of the same material has an external diameter of about 2½ inches. Tubing is also sized by its inside diameter. Therefore, a 1-inch tube has an inside diameter of 1 inch and, since its walls are thin, its outside diameter is only about 1⅛ inch.

For this reason, a length of 1-inch galvanized pipe looks and is thicker than a length of 1-inch tubing, though both the tube and the pipe are considered to be the same size.

At the same time, the copper tubing has approximately one pipe-size greater water-carrying capacity. Where you would use 1-inch galvanized pipe you can use ¾-inch copper tubing and get the same results. This relation holds true for the entire range of pipe sizes ordinarily used in a pri-

vate home. The reason is the much smoother inside surface of the copper tube. The zinc with
which the iron pipe is coated (galvanized) is
rough enough to slow the passage of water.

FITTINGS

Every time you need to join two or more pipes
to one another you have to use a fitting. Every
time your pipe or your copper tubing must take a
tight turn, you have to cut the pipe and insert a
fitting.

Fitting descriptions—Fittings are described by
the material of their manufacture, for example
galvanized iron fittings, plastic fittings, and so on.
Fittings are further described by the size of the
pipes or tubes they accommodate. A 1-inch
fitting "takes" 1-inch pipe, though the holes for
the pipe have a diameter of 1⅜ inches and the
fitting itself may be much more than 2 inches
across.

In addition to material of construction and
size, fittings are further described by the method
used to join them to the pipe(s) and tube(s) and
their function. For example, if the fitting was
made to accept threaded pipe it would be called a
threaded fitting. If the fitting was made to lead
the pipe around a corner (that is, to join two
pipes at an angle), the fitting would be called an
elbow (or bend or sweep) and its degree of turn

would be specified. If the elbow made a sharp
right turn it would be called a 90-degree elbow.

Thus, if you want to join two 1-inch galvanized
pipes at a right angle you need a 1-inch, 90-de-
gree, galvanized iron elbow. Should you want to
join the ends of two ½-inch copper tubes at a
45-degree angle, you need a ½-inch, 45-degree
copper sweat fitting, because copper tubes are not
fastened by screw threads, as are galvanized pipe,
but are sweated (soldered) together.

Here is a description of the more common
fittings used in plumbing.

Couplings are short, straight lengths of tubing
or pipe. They are used to connect the ends of two
tubes or pipes without introducing a bend. The
coupling used for copper is smooth on the inside
except for a circumferential indentation at its
center: this is there to prevent either tube from
being inserted more than halfway. Copper tube
couplings are sweated (soldered) in place. Couplings for use with galvanized pipe are internally
threaded (female threads) at both ends. The
externally threaded (male) pipe ends are screwed
into place and the joint is made.

Elbows are couplings that join two pipes at an
angle. They are manufactured for all kinds of
pipes and pipe sizes and in a variety of angles.

Tee fittings join three pipe (or tube) ends at
right angles.

Wye fittings are similar to Tee fittings except

7. Just a few of the many galvanized iron pipe fittings currently stocked by plumbing
shops. Left to right: A Tee, a street elbow, a 35-degree elbow, a 90-degree elbow and a
union. All of these fittings are made for use with threaded pipe.

8. A small sampling of the many copper "sweat" fittings manufactured. Fittings also having threads are adapters and are used to connect copper tubing to threaded pipe.

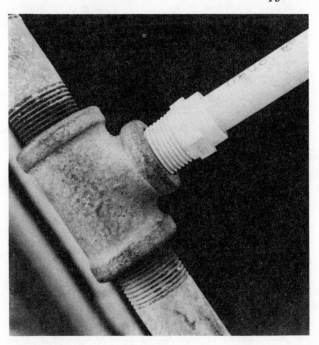

10. An adapter connecting plastic pipe to galvanized pipe. The adapter must be screwed into the Tee before the plastic pipe is solvent-cemented in place.

9. Left to right: An elbow adapter to connect copper tubing to a male-threaded pipe end. An adapter to connect tubing to female-threaded pipe or fitting. An adapter to connect male-threaded pipe to tubing.

that the three-way joint doesn't form a right angle. Whereas the Tee fitting looks somewhat like the letter *T*, the Wye looks somewhat like the letter *Y*.

Double Wye fittings join four pipe ends at angles other than right angles.

Cross fittings connect four pipe ends at right angles.

Adapters (sometimes called transition fittings) are used to join two different kinds of pipe together, or sometimes similar pipe that is joined by two different means. For example, an adapter is used to connect the end of copper tubing to galvanized pipe. An adapter is used to connect plastic pipe to thin-wall brass pipe.

Nipples are short lengths of pipe externally (male) threaded at both ends. Some are so short that the thread at one end runs into the thread at the other. This type is called a close nipple.

Pipe extensions are a form of coupling, but differ in that they have a female thread at one end and a male thread at the other.

Reducers are used to couple the ends of two pipes of differing diameters together.

11. Left to right: A reducing coupling used for connecting the male-threaded ends of two different sizes of galvanized pipe. A bushing, which permits a smaller-size threaded pipe to be connected to a larger-size threaded hole. A coupling used to connect two equal-size lengths of threaded pipe. A plug used to close a threaded hole.

Bushings are also used to connect two pipes of differing diameters together, but more often to fit a threaded pipe end into an internally threaded, but oversize hole. For example, to fit ¾-inch pipe into a hole threaded for 1½-inch pipe, a bushing is screwed onto the end of the pipe, then both are screwed into the hole.

Unions are used to connect pipe ends together in the same manner as couplings. The important difference between unions and couplings is that unions can be taken apart without unscrewing the pipe. This is a very important advantage; therefore, unions should always be incorporated whenever more than a few lengths of threaded pipe have to be joined. For example: assume that you are going to connect a top-floor lavatory to a basement water supply. You cut the line and insert a Tee. Then you "make up" the pipe, section by section, fitting after fitting until you reach the top floor. All in all you may have screwed 10 lengths of pipe and 9 fittings into place before connecting the lavatory faucet. You turn the water on and find that the basement Tee leaks. You tighten the joint as much as you can, but it still leaks. The Tee's thread is defective. To get at the Tee you have to unscrew every single pipe and fitting that follows. If you had installed a union after the Tee, you could "break" the line there.

Plugs are used to close internally threaded holes—for example, the second hole on a radiator drilled for a right-hand and a left-hand connection, only one of which is used.

Caps are used to close the ends of externally threaded pipe ends.

Compression fittings (and joints) depend on the compression of a metal or flexible ring to form a watertight seal between the pipe end and

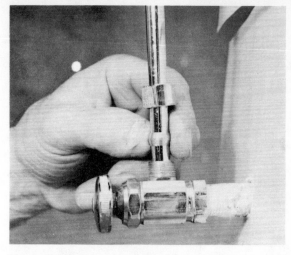

12. A metal-ring compression fitting (joint) used to connect a water-supply tube to an angle shutoff valve. When the metal ring is slid down and the nut is tightened a watertight joint is made, which can be disassembled and reassembled if the nut isn't overtightened and the tube isn't replaced in a different position.

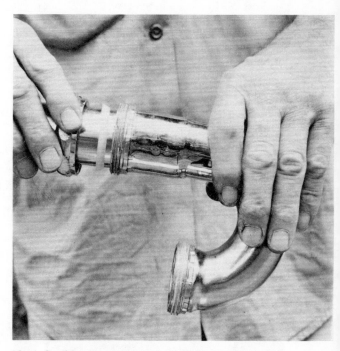

13. A flexible-ring compression fitting (joint) used to join a trap to a thin-wall brass tail-pipe (which connects to a sink or a lavatory). When the flexible ring (white plastic in this case) is slid down and the nut is tightened a watertight joint is made. The joint can be opened and closed innumerable times without difficulty.

the fitting. When flexible rings are used the joint can be readily disassembled and reassembled without leaking. In some designs the depth to which the pipe or tube enters the fitting can be varied at will (this type is called a **slip fitting**). When a metal ring is used, a great deal of care must be exercised in assembling and later in reassembling the joint if it is not to leak when closed the second time.

Flare fittings require a special tool which "flares" the end of the copper or plastic tube. The fitting's flange nut compresses the flared tube end over a conforming part in the fitting. Flared fittings, like compression fittings, can be disassembled and reassembled without disturbing the rest of the piping.

FIXTURES AND APPLIANCES

In plumbing parlance toilets, lavatories, sinks and tubs are all called fixtures. Dishwashers, clothes washers and the like are considered appliances.

VALVES AND FAUCETS

A valve is a device for controlling and stopping the flow of water and of gas (for heating). Valves are used when necessary somewhere along a run of pipe. When a valve is used at the end of the pipe run it is called a faucet. Essentially, this is the major difference, although there are many valve and faucet designs. Valves are sometimes named after a specific use, such as curb stop, the valve in the water service line generally positioned underground near the curb. Shutoff valves are used in line with fixture water supplies, not to vary the flow, but to shut it off.

As stated, the valve at the end of a pipe run—the valve that is designed to be in daily use—is called a faucet. Its function is identical to that of a valve, but its construction often differs. Essentially a faucet is designed for frequent use in partially opened as well as fully opened and closed positions. And faucets are designed to be repaired; many valves are not. Faucets are also called taps, pet cocks and sill cocks.

CHAPTER THREE

Tools

All you really need on hand for emergencies are a few tools. When you undertake a plumbing expansion or remodeling project you can purchase the less expensive tools you may need and rent the others. There is no need to have them all at home.

The following are the most-used tools—those that are advisable to purchase and keep in your workshop.

Screwdrivers
Pliers
Small Stillson wrench
Medium-size crescent wrench
Valve-seat removal tool
Force pump
Closet auger
Flashlight

In addition, purchase some string packing and a box of assorted washers and associated screws (they come in the box).

Screwdrivers—A 6-inch and a 10-inch standard screwdriver, plus a 6-inch and a 10-inch Phillips-head screwdriver should take care of all the screws you will encounter.

Pliers—The most useful pliers and the tool you will probably use most of all is **channel lock,** the name of an offset pair of gas pliers that cannot slip apart under stress. They cost much more than the usual gas pliers, but are worth it. The 10-inch size is perhaps best for plumbing.

Stillson wrench—You will need two if you have to disassemble stubborn pipes and unions, but for most repair work a single 10-inch wrench with an offset head (new design) will do fine.

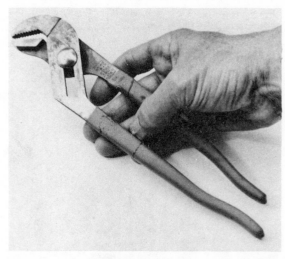

14. Next to a screwdriver, a pair of channel-lock pliers is probably the most useful tool you can have in your kit.

15. A small, offset Stillson wrench. The offset type, with its handle at an angle to its jaws, is more adaptable to tight quarters than a standard wrench.

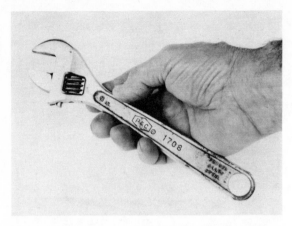

16. A small crescent or adjustable wrench is a poor man's answer to a set of wrenches—not as good, but good enough for most jobs.

18. A plumber's friend is a friend indeed. The working end of the better type of plunger.

17. A valve-seat removal tool with a valve seat on its end. The straight-bar type, which requires the use of a wrench, works better than the angle type, which needs no wrench.

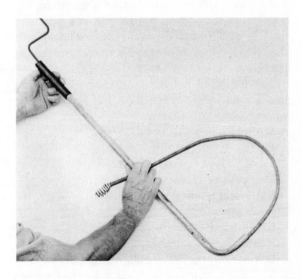

19. A closet auger. The coiled end is directed into the water closet (toilet bowl) while the handle is turned.

Crescent wrench—Although not as good as a single-size open-end wrench, the crescent will handle a range of sizes and is therefore considerably less expensive than a set of wrenches. The 10-inch wrench will be most useful.

Valve-seat removal tool—This is a bar of steel about 1 foot long having a tapered square end and a tapered octagonal end. It is used in conjunction with a crescent wrench to remove valve seats.

Force pump—Sometimes called a plumber's friend, the force pump is used to force drain obstructions down and out of waste and soil lines. Buy the type with the adjustable lip—the sturdier the better. The simple, soft rubber hemisphere type isn't half as effective.

Closet auger—A snake with a handle, which is convenient. The snake end of the flexible metal cable is directed into the toilet bowl while the handle is turned. With luck and continued effort it is possible to either push the obstruction down and out of the toilet's trap, or to hook onto the obstruction and pull it out. Purchase the most sturdy design you can find. If the snake is soft and limp the device is useless.

Flashlight—An ordinary, mundane aid, but very useful nonetheless. Get one with a magnet on the handle. It can be very useful.

ADDITIONAL TOOLS

Following is a short list of tools you may wish to invest in before you have need for them, since they are useful in emergencies.

Large snake—This is necessary for clearing stoppages that cannot be dislodged by chemicals or the short closet auger. The auger has about a 4-foot reach, while a large snake is 25 feet or more in length.

There are two types that are commonly sold. One has a flat cross-section and is the least expensive. The other snake has a circular cross-section and is composed of steel wire tightly twisted. The flat strip type tends to fold and twist up when the going gets tough, so that its usefulness is limited. The other type is best, but you have to purchase a snake that is close to ½ inch in diameter. Anything less and it too is almost worthless.

Large Stillson wrench—To open a threaded joint without a vise, or when you cannot place the pipe or fitting in a vise, you need two wrenches: one to hold the pipe or fitting, the other to turn the second portion of the joint. As a second Stillson wrench, an 18-inch wrench is about right. It will handle up to 2-inch pipe.

Propane torch—The torch is necessary for opening and remaking sweated joints in tubing, and for closing leaking sweated joints and acci-

20. Tools and material for soldering: propane bottle filled with liquefied gas, torch, solder and soldering paste.

dental nail holes. The torch is also used for loosening stubborn threaded joints.

Don't buy the foreign GAZ torch even though it is half the price of the usual American torch, because the GAZ torch can only be used with that company's replacement cylinders. Many companies make the propane cylinders that fit the standard American propane torch. You can buy the American metal bottle of gas in almost any hardware and sporting goods store. When you get the torch, purchase some solder and solder paste (flux) at the same time.

STILL MORE TOOLS

The special tools, those seldom used for repair work, are discussed along with the pipe and fixtures they are used with. For example, thread-cutting equipment, how and when to use it, is covered in the following chapter, Working With Threaded Pipe, and so on.

CHAPTER FOUR

Working With Threaded Pipe

TYPES THAT ARE THREADED

All the metal pipe used for plumbing can be threaded and some types almost always are. Galvanized wrought iron, black iron (wrought iron without its coating of zinc), copper and brass pipe are almost always joined by threaded joints. Cast iron is almost always joined by means of lead caulking. This is a process wherein oakum (rope) followed by lead is pounded into the space between the pipe and its fitting to make a watertight joint. However, some cast-iron fittings have threaded openings that accept threaded pipe.

Copper tubing is never threaded, but thin-wall brass sometimes is. Thin-wall brass is a large diameter (1¼-inch or more) pipe that is used as a drainpipe immediately beneath sinks, lavatories and tubs. Most often thin-wall brass pipe is joined to its fittings by means of a flexible-ring compression joint. But sometimes, where there is more than a single pipe as in the case of a double sink, one thin-wall pipe will be factory threaded, as will one opening on an associated drain fitting. The purpose is to provide a stiffer joint than is possible with a flexible ring. Factory thread is cut at the factory. It is not done in the field or at a plumbing supply house, and the thread is a fine machine thread, not a plumbing thread.

THREADS

If you have been working with nuts and bolts to any extent you know that two threads are commonly used. One is called fine thread and the other coarse. A fine-threaded bolt has more threads per inch than does a coarse-threaded bolt. If you look closely you can readily see the difference. A fine-threaded nut will not run on a coarse-threaded bolt and vice versa.

Three threads are used in plumbing. The aforementioned machine thread, which is used only on thin-wall brass pipe, water hose thread and standard pipe thread. Water hose or garden hose thread is very coarse. It is used only for garden hose fittings and on the ends of faucets to which garden hoses will be attached.

All other pipe thread is standard plumbing thread. All pipe fittings have matching thread and all pipes are manufactured with the same thread. It is almost always right-handed, but some fittings can be had with left-handed thread.

Standard plumbing thread can be cut with portable tools and some plumbing supply shops will cut pipe to sketch, which means they will cut and thread pipe to specified size. When you purchase a length of pipe, its ends are precut with standard, right-hand plumbing thread. When you have a shop cut thread for you, you get the same standard thread.

So far, simple enough, but there is one aspect of plumbing thread that is important to understand and appreciate. That is, pipe thread is cut at an angle. If you examine the threaded end of a pipe carefully you will notice that the diameter of the pipe at its ends is somewhat less than it is farther back. The difference is about five degrees, which makes the pipe end a tapered, threaded plug and not a bolt. The tapered plug can enter the female-threaded hole so far and no farther. Tightening the plug into its hole locks the pipe in

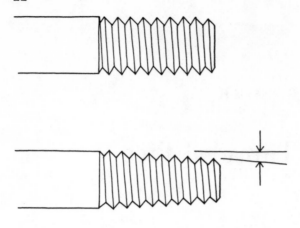

21. Pipe thread is pitched. As a result, the pipe can be locked into a fitting by tightening; machine thread cannot. Pipe-threaded fittings can be damaged by overtightening, which expands the fitting beyond normal "spring" return.

place and *seals the joint*. If you overtighten the joint, the joint will continue to hold water, but if you should back the pipe off just a little way, it will leak badly. Many times, tightening the joint up a second time doesn't make it watertight. The reason is that forcing the hole beyond its normal "give" permanently distorts it. The sides of the plug and enveloping cone are no longer parallel and therefore will not hold water.

ASSEMBLING A THREADED JOINT

The male-threaded end of the pipe is covered with a layer of pipe dope, which is a thick, paintlike mixture sometimes called pipe joint compound. As an alternative you can rub pipe dope in stick form over the threads, or you can use a layer of special plastic tape made for the purpose. Of the three, the compound is by far the best.

22. You can hold pipe in a standard vise by making a jig such as shown from scrap lumber. Always "dope" the male end of threaded pipe before assembling the joint.

23. You need a Stillson wrench to "make up" the joint. Just be careful not to over-tighten. If you do not have a vise you need two wrenches to make up or break down a screw-thread pipe joint.

It is very important that pipe dope always be used in the assembly of a threaded joint. The dope does many things. It seals the joint: very often the same joint made up "dry" will leak. The dope lubricates the threads so that the parts are more easily assembled. The dope prevents the threads from heating up and stopping the pipe from being turned before the joint is properly tightened. The dope also enables you or your heirs to disassemble the joint sometime in the future. Without dope an iron-to-iron joint can rust permanently solid.

After the dope is applied the pipe is lined up and turned in its fitting as far as it will go by hand. Then it is made fairly snug, but *never* over-tightened. It is far better to go easy on the wrench handle than to give it all you have. If the threads have been properly cut, if the parts are not cross-threaded, meaning the pipe has not been started at an angle to the fitting, and if pipe dope has been used, all you need to do is make the joint a little more than hand tight to hold pressure. Should the joint leak when you let the water in, you can always tighten it up then.

MEASURING PIPE

As far as plumbing is concerned, a pipe has only two dimensions: size and length. Pipe size is based on its inside diameter. If you have a length of pipe handy, put a ruler across one end and measure its inside diameter. The largest number nearest the actual inside diameter is the pipe's size. It is called nominal size because the figure is not exact. For example, the inside diameter of ¾-inch pipe is actually 0.82 inch, which is not three-quarters of an inch.

To find the "pipe" size of a fitting end that has female threads, screw a section of pipe into the fitting. The size of that particular end of the fitting is the size of the pipe it accepts.

Now to length. This is not what it appears to be because the pipe's ends enter a fitting and/or a valve or a faucet for a short distance. Therefore the actual length of a section of pipe connecting two fittings or a fitting and a faucet is always the distance between the facing edges of the two fittings *plus* the distance the pipe's ends enter the fittings.

There are several methods used for establishing necessary pipe length. The experts have a table that gives them all the dimensions of the fittings, valves and faucets they may want to connect. With this data they can figure necessary pipe length from one fitting's center to the next. This is the method used to cut pipe to sketch—that is to say, to cut pipe by examining a blueprint of the building. A much simpler method, which does not require fitting data, consists of actually measuring the separation desired between two fittings and then adding the two distances the pipe enters the two fittings.

For example, assume that you want to connect a faucet to an existing pipe. You know where you want your faucet and you know where the pipe is. The first step would be to cut the pipe and insert a Tee fitting. (Cutting and threading are discussed a few pages on. Right now we are concentrating on measuring necessary pipe length.) The second step would be to temporarily position your faucet, and for simplicity we will imagine using a simple, old-fashioned, screw-on faucet. The distance between the end of the female opening on the Tee and the female opening on the end of the faucet is measured. Let us say the distance is exactly 2 feet. And, let us say, we are running ¾-inch pipe. Referring to the accompanying table, we find that ¾-inch pipe enters a fitting ½ inch. Thus we need 2 feet, 1 inch of pipe to do the job.

If the table isn't handy, just remember that the most commonly used pipe sizes, ½-, ¾- and 1-inch, enter fittings ½, ½ and ¾ inch respec-

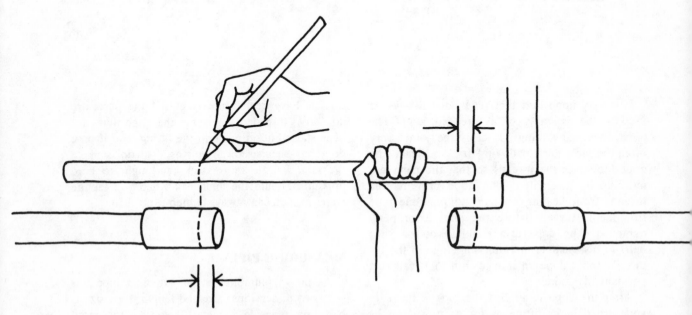

24. One of several ways of determining the length of pipe necessary to join two fittings or a fitting and a valve. Method shown is the easiest and least susceptible to error.

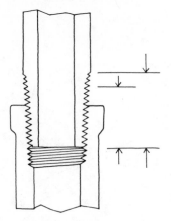

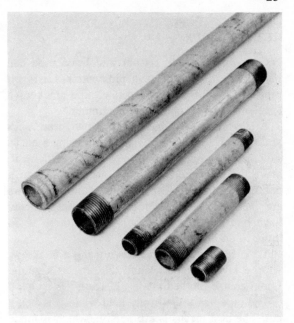

25. Cutaway view of threaded pipe. Note that the female fitting is never tightened all the way up on the male pipe end.

26. An unthreaded length of pipe followed by an assortment of nipples. The one at the extreme right is called a close nipple. It is almost all thread and permits the two fittings that it connects to touch. Nipples are available in a variety of lengths, so that by using a number of them plus couplings it is possible in many instances to avoid the need for cutting thread.

tively. Always remember that a pipe has two ends, therefore the figures given must be doubled —one for each end.

In some instances it may be difficult to hold a fitting in position while you measure the distance from its end to a second fitting. In such cases, connect one fitting to the section of pipe and hold the pipe up against the side of the second fitting. Then you can mark the cut on the pipe.

The distance a pipe enters a fitting is not critical. You can vary it a bit and still have a watertight joint. Just start out by hand-tightening the joint. Then, if you have too much pipe, you can remove it, cut it shorter, rethread it and try again. If you attempt to force the pipe into the fitting to rid yourself of the excess pipe and you cannot bring the parts into position, there is a good chance that you will have to replace the forced fitting, as it may not hold water afterward.

PRECUT, THREADED PIPE

In addition to full-length sections of pipe, which are generally 20 feet long and always come with one free coupling on their ends, many plumbing supply shops and hardware stores now carry pipe in various lengths, threaded at both ends. And, nipples from close (3–4 inches long)

to 12 inches long have always been available. Thus, with a little judicious planning and a little compromising it is often possible to install pipe without cutting and threading.

If you cannot make do with what is available, there are plumbing supply houses that will cut pipe to your order at so much per cut and threading. If all you require is a few lengths, this is your best deal. If you need more than a few pieces, or if there are no shops providing this service nearby, you will have to cut and thread it yourself. This is neither complicated nor difficult, but it does require equipment that is comparatively inexpensive if you are doing an entire house, but very expensive for just a few pieces of pipe. The alternative, not always possible, is to rent the necessary tools.

CUTTING PIPE

Galvanized, black, copper and brass pipe can be cut with a hacksaw or a pipe cutter. Cast-iron pipe can be cut with a hacksaw, pipe cracker (a special tool) or a chisel and a hammer.

Attacking with a hacksaw has two drawbacks and two advantages. It is hard work using a hacksaw on a large-diameter pipe, and it is difficult to make the cut square, which is necessary to do a good threading job. On the other hand, hacksaws are inexpensive, most home workshops already have one and, once the pipe is cut, there is no more to do before threading.

To ensure a square cut when using a hacksaw, wrap a layer of tape around the pipe as a guide. Use a coarse-toothed blade (18 teeth per inch).

The pipe has to be firmly held while it is being cut. The best means is a pipe vise, which is made especially for holding round objects. Lacking a pipe vise, you can use an ordinary machinist's vise if it is large enough for the pipe that is to be held. If the size relation is correct you can sometimes "jam" the pipe below the flats on the vise jaws. If not, you can use some wood scraps placed between the flats of the facing jaws and the sides of the pipe to provide a "bite" on the pipe without flattening it. The arrangement is certainly not as convenient as a pipe vise, but it will work.

The pipe cutter is far easier to use and much faster. It always makes a perfectly square cut. A pipe cutter is a sort of clamp with two wheels that roll on the pipe and a third wheel, ground to a sharp edge, that does the cutting. The edge of the wheel is positioned on the desired line of cut. Then the handle of the device is turned until the cutting wheel presses lightly into the metal. Next the entire gadget is swung around the pipe. This cuts a small groove circumferentially into the pipe. Now the handle is given another partial turn, and once again the cutting wheel presses into the metal. The cutter is rotated around the pipe again: the groove is now deeper. The process of tightening up on the cutting wheel and rotating the cutter is repeated again and again until the pipe is severed.

In severing the pipe—there is really very little

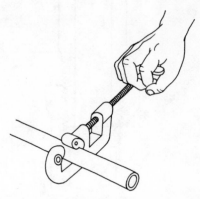

27. To use this pipe cutter, its handle is given a partial turn every time the tool is rotated around the pipe.

cutting action—a lip is raised on the outside surface of the pipe end and a burr around the inside end of the pipe. The height of the lip and the burr vary depending on how sharp the cutting wheel is and how much "cut" you try to take with each rotation, but they are never absent. The burr must always be removed, as it severely reduces the effective inside diameter of the pipe. The external lip can be ignored in many instances if it isn't too high. If the die (see next section) passes over it, the lip can be ignored.

The internal burr can be removed with a rattail file, but more rapidly and easily with a plumber's reamer held in either a carpenter's brace or a plumber's ratchet handle. A flat file is used on the external lip.

THREADING PIPE

You need a pipe-thread die of a size that matches the pipe's size, or an adjustable **die** that can be matched to the pipe, plus its handle, called a **stock** (stock and die). And, of course, you need the all-important vise to hold the pipe.

The larger opening on the die is slipped over the pipe end and pressed firmly against it. It is difficult to do this incorrectly, as the die will not slip over the pipe end if you attempt to start it the wrong way. The die, firmly pressed against the pipe end, is turned clockwise—to your right as you face the pipe end—about one turn, then backward about one quarter turn. Cutting oil is

28. Using a stock and die to cut pipe thread. The die is pressed against the end of the pipe as shown. Then the handle is turned clockwise until two threads appear beyond the die. Cutting too much thread results in a leaky joint.

29. Keep squirting cutting oil—any oil will do if you do not have thread-cutting oil—as you cut the threads. The oil cools the metal and makes for a better, cleaner cut.

now squirted into the die and onto the pipe end. Then the die is again given one turn forward and a quarter turn back. This is done to clear the chips. The forward and backward turning and intermittent oiling is continued until the end of the pipe is either flush with the outer surface of the die, or two or three threads protrude.

There are two important points that need explaining, and this will be done before we go further.

First, the die must be lubricated. If it is not, the die wears rapidly, considerably more than normal pressure is required on the stock and, most important, the threads are ragged and unsatisfactory. Thread the pipe without lubricant and most likely the joint made with that pipe end will leak no matter how much pipe dope is used and no matter how tightly the joint is made up.

Use cutting oil especially formulated for the purpose and available at plumbing supply houses, or use high-detergent motor oil, or use a combination of kerosene and motor oil, but use a lubricant.

The second point is, where do you stop cutting thread? If the die itself carries instructions, fine. If the man who rents or sells the stock and die knows, no problem. But if you cannot secure firm advice you have to use the following guidelines, because if you do not run up enough thread the joint will be weak and it will leak. If you run up too much thread the joint will be very strong, but it will leak.

The number of threads you should run up isn't fixed; you can have a couple more or less. When you vary from the desired number more than that, however, the joint will be defective.

Very simply: Run the die up on the end of the pipe until the pipe end is flush with the die surface nearest yourself. Spin the die in the reverse direction and back it off. Next place a ruler against the thread you have cut. Measure the length of the threaded section. Refer to the accompanying table. If there isn't sufficient thread, replace the die, add lubricant and cut some more. Remove the die and measure again. When you have cut sufficient thread, run the die back up the pipe end. Note how much pipe extends beyond the die. This will be your guide for future thread cutting.

Typically, ½-inch pipe requires ¾ inch of thread
¾-inch pipe requires ⅞ inch of thread
1-inch pipe requires 1 inch of thread

As a secondary check, dope the pipe end and run it into a fitting. Mark the edge of the fitting against the pipe end. Remove the pipe and measure how far the pipe entered the fitting when pulled up snug.

Typically, ½-inch pipe will enter a fitting ½ inch
¾-inch pipe will enter a fitting ½ inch
1-inch pipe will enter a fitting 1 inch

FEMALE THREAD

Female thread is cut with a tool called a **tap.** It looks very much like a tapered bolt with four longitudinal slots. The only time female thread is cut on the job in plumbing is when a very small diameter pipe is to be connected directly to a much-larger-diameter pipe. In such cases a hole is drilled in the larger pipe. The hole is a little smaller than the tap that is to be used. After the hole is drilled it should be reamed with a reamer having a taper of ¾ inch to the foot, which is the taper of pipe thread. The reaming isn't too important, but if you are making a hole in cast-iron pipe, which has a thick wall, or in large-diameter galvanized, which also has a thick wall, it is best to ream it.

The reamer and the inner edges of the hole are lubricated with cutting oil. The reamer is inserted, given one full turn to the right and then backed off. This is repeated until the tap is fully into the metal *but not all the way through.* If you run the tap completely through, the hole and thread will be too large. To cut thread for pipe, a pipe-thread tap must be used. The standard machine-thread tap will not do.

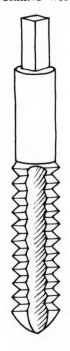

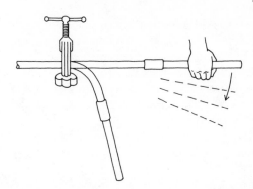

30. Typical tap used for cutting pipe threads.

31. When bending pipe, do not try to make the entire bend at one point on the pipe. Instead, bend the pipe a little, then move the pipe a fraction of an inch out of the vise and bend it a little more. Do this until the bend is completed. If you don't, you will kink the pipe. If you cannot bend the pipe at all, extend it as shown.

DRILL SIZES TO BE USED WITH PIPE TAPS

Pipe size	Threads per inch	Drill size
⅜	18	$1\frac{9}{32}$
½	14	$2\frac{3}{32}$
¾	14	$1\frac{5}{16}$
1	11½	$2\frac{4}{32}$

BENDING PIPE

Pipe is not normally bent; that is to say, the job is usually laid out so that there is no need for bending. However, when plans go awry or when a fitting with the particular angle required is not to be had, pipe is bent. It does it no harm, it is just difficult to do without proper equipment.

To bend pipe without a hydraulic ram and guide, place the pipe in a vise (extend it if you have to by adding a coupling and another length of pipe). Bend by pulling on the pipe end just a little, but enough to introduce a tiny bend. Move the pipe in the vise a fraction of an inch; bend again; move again. In effect you are making a large circular bend by taking a lot of little bites. Do not attempt to make the bend at one point; you will flatten the pipe. This may sound Herculean, but it is actually easy with a sufficiently long pipe.

If one end of the pipe is attached, you can bend the pipe slightly by using a large Stillson wrench. The wrench is tightened onto the pipe and then pulled toward the pipe. This trick is useful to get a little offset into a pipe that doesn't hit its mark. But remember that all bends and offsets act to shorten pipe length.

CHAPTER FIVE

Working With Copper Tubing

Although considerably more expensive than galvanized pipe, copper tubing is almost always used in new building (unless local plumbing codes permit the use of plastic pipe). There are many reasons for selecting copper over the less expensive galvanized. Copper tubing is lighter in weight; this makes it easier to carry and position. Size for size, copper tubing has a smaller outside diameter, permitting smaller holes and less clearance for fittings and joints.

Copper is easily bent, so many joints and fittings required with galvanized are not needed for the same run of copper. Copper is more easily and quickly cut and joined than galvanized. And the tools needed for cutting and joining copper cost perhaps as little as one-tenth that of the tools needed to cut and thread galvanized.

In many soils, copper is more corrosion resistant than galvanized. It is therefore the better choice of metal for use in the water service line, which runs underground.

Copper, however, has one drawback in comparison to galvanized: it is soft and must be protected from foot traffic and nails. You just cannot lay copper tubing across a cellar floor and forget it, the way you can galvanized. Copper has to be shielded by a layer of concrete or wood. And copper cannot be run through a wall without strips of metal to protect it in areas where shelf- and picture-supporting nails may be driven. The nails will enter the copper tubing without the slightest resistance.

SIZE COMPARISON

A portion of the cost difference between copper tubing and galvanized pipe is reduced by copper's greater water-carrying ability. While galvanized becomes rough with time, copper retains its inner smoothness. Copper, therefore, can always be used one size smaller than that which would be used if the run was made of galvanized. Thus, if you are replacing a 1-inch galvanized pipe you can install a ¾-inch copper tube without loss of water flow. Reducing pipe or tube size also reduces the cost of the associated fittings.

TYPES OF TUBING

There are two types of copper tubing manufactured for plumbing: rigid and soft. The rigid is available in straight lengths of 20 feet. It is used where the pipe may be visible and a neat job is desired. Rigid tubing is normally not bent on the job, though the tubing can be bent; instead, elbows are used. Soft tubing is manufactured in straight lengths of 20 feet and coils of 60 and 100 feet. Soft tubing is more easily bent and is always used when and where the tube has to be guided through walls and around turns. The coiled tubing is simply uncoiled for use. However, no matter how carefully you uncoil it, a few unimportant kinks are always visible. Many plumbing supply shops and hardware shops will cut both the coiled tube and the straight tubing to

whatever length you may require, though the cost
per foot is higher than if you do it yourself.

Wall thickness—In addition to the basic divi-
sion—soft and rigid—copper tubing is also manufactured in a number of wall thicknesses desig-
nated K, L, M and DWV. Type K is color-coded
green. It has the thickest wall, is the strongest
and is the most expensive. Type L is color-coded
blue. It has medium-thick walls and is the type
most often used for home plumbing systems.
Type M is color-coded red. It has the thinnest
walls of all the types and is seldom used for anything else but radiators and cooling coils. The let-
ters DWV stand for drain, waste and vent; this
type of tube is made especially for drainpipe
work and is manufactured in rigid form only. It is
somewhat thinner than type L, which makes it
cheaper than type L in equal diameters. So when
you are running drain-, waste or vent pipe, your
best choice is DWV.

TUBING SIZE

Tubing is sized exactly the same as galvanized
and other pipe. The size of the tubing is based on
its inside diameter, even though the inside diameters of the various types of tubing vary considerably. For example, the 1-inch K-type tube actually has an inside diameter of .995 inch, while
the 1-inch type-L tube has an inside diameter of
1.025 inch.

MEASURING TUBE LENGTH REQUIRED

It is easier to measure the length of tubing you
may need to connect two fittings a specific distance apart than it is to do the same for galvanized or other threaded pipe. The reason is that
the tubing end can be slipped into the fitting for a
quick check, whereas you have to thread the pipe
to do the same. It is also much easier to shorten a
tube that you have tested and found to be too
long than to shorten a pipe that has been tested
and found long.

Measuring between fittings—Position the fittings where you want them. Measure the distance between the two facing edges. Add the two
distances the tube will enter the two fittings (see
illustration 24). Use the following table as a
guide.

DISTANCE TUBING ENTERS A FITTING

Tubing size	Entering distance
⅜	⁵⁄₁₆
½	½
¾	¾
1	¾
1¼	⅞
1½	1

An alternative approach consists of positioning
the two fittings as before, then inserting the end
of the tube into one fitting and holding the tube
up against the side of the second fitting. The distance the tube will enter the fitting plus a half
inch or so is marked on the tube. It is then cut
and the fittings are slipped over the tube ends.
The separation distance is measured and corrected, if necessary by cutting a piece off the end
of the tube.

Measuring around a turn—The easy way to
measure around a turn is to temporarily position
one tube end in one fitting, bend the tube around
the corner (bending is discussed later in this
chapter) and then position the second fitting
against the tubing, mark it and cut it a little
longer than necessary. There are formulas that
enable one to compute tube length necessary to
go around a curve. They are based on radius,
circumference and pipe diameter, but aren't necessary if you allow plenty of tubing for the bend
and cut the tube after it is bent.

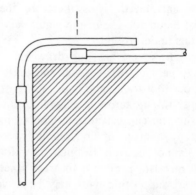

32. When you have to bring your pipe around a
corner and connect it, make the bend first, then cut
it to fit.

CUTTING TUBING

Tubing can be cut with a hacksaw or a tubing cutter. If you have but a few cuts to make and have a hacksaw, there is no real need for a cutter. But if you have lots of cutting to do, it is wise to invest in a tube cutter. It works much more quickly, easily and accurately than a hacksaw, and the tube cutter is not very expensive.

Sawing tubing—A hacksaw with a fine-toothed blade (32 teeth to the inch) is best. The tubing is held by hand in a simple jig made of a wood cleat nailed to a bench top or nailed to a 5- × 6-inch piece of wood which can be nailed to a bench top or held in a vise (see illustration 127). The cut is made as squarely across the tube as

possible. This is fairly easy to do on the small-diameter tubing, but increasingly difficult on larger-diameter tubing. Should you need to cut tubing larger than ¾ inch, make yourself a miter box of wood scraps to guide the saw.

A reamer or a rattail file is used to remove the internal burr, and if there is any, the external burr. It is very important that the internal burr be removed, especially on small-diameter tubing, as the burr will severely reduce the rate of water flow.

Using a tubing cutter—In principle and operation a tubing cutter is exactly like a pipe cutter. The only difference is size and strength. The cutter doesn't need to be as strong for copper tubing as it does for iron pipe.

33. Adjust the cutter's handle until the cutting wheel rests firmly on the line of cut. Then rotate the device around the tube. Give the handle another turn and rotate the tool again. Repeat until you have severed the tubing.

The cutter's handle is adjusted until the cutting wheel rests firmly on the line of cut. Then the entire device is rotated around the tube. The cutting wheel lightly grooves the tube as the tool is rotated. Then the tube cutter's handle is given another turn, more pressure is applied to the groove via the cutting wheel, and the tool is again rotated around the tube. These actions are repeated until the tube is severed. Usually, half a dozen rotations of the cutter around the tube do the job. Avoid applying too much pressure on the cutting wheel. Doing so will distort the tube.

After the tube has been cut the inner burr must be removed. Generally the tubing cutter has a small reamer attached. If it hasn't, use a rattail file or a plumber's or carpenter's reamer.

JOINING TUBING

Copper tubing can be joined to other copper tubing, fittings, or valves and faucets by sweating (soldering), compression joints and flare joints. Each method has its advantages and disadvantages. Sweated joints are easy to make, fast, least expensive in terms of required fittings and permanent. However, as you need a propane torch to make them, they cannot be made near flammable material and they also require a propane torch for disassembly.

Compression joints are fast and easy to make. The fittings needed are more expensive than those necessary for sweating, and while the joint is permanent it is not as strong mechanically. However, on the plus side, if a flexible ring is used the joint can readily be opened and closed. If a metal compression ring is used the joint can only be opened and closed with extreme care. If the parts are not properly aligned the joint will leak when closed a second time.

A flare joint requires a lot more care to make properly than the other joints. It can be opened and closed innumerable times, but the necessary fittings are much more costly than those needed for sweated and compression joints.

COPPER TUBE FITTINGS

To a fair extent fittings for copper tubing duplicate what is manufactured for galvanized iron.

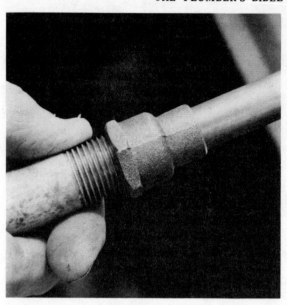

34. An adapter is used to connect a threaded pipe to the end of a copper tube. It is best to solder the fitting to the tube first.

There are copper couplings, Tees, Wyes, elbows, reducers, adapters and so on, plus one fitting that is not available for threaded pipe: the slip fitting. It is similar to a coupling in that it connects two in-line, similar-diameter tubes, but it differs in one important way. A slip fitting can be slipped any distance up a length of tube, then slipped back over the pipe ends and sweated in place. This enables you to cut a middle section out of a tube and replace it with two short tubes and a Tee or a Wye, and is very useful when you are connecting new tubing onto an existing line.

The greatest variety of fittings will be found among those designed for sweating, because tubing is far more often joined by sweating than by any other method. However, there are a sufficient number of different fittings normally manufactured for compression and flare joints to take care of almost all pipe run situations.

SWEATING A JOINT

Sweating (soldering) is rapid, certain and easy once you understand the principles involved and follow certain simple rules.

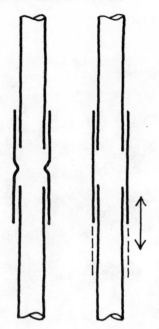

35. A slip fitting has no internal protrusion and can therefore be slipped along a tube as far as necessary and back again to complete the joint. A coupling has an internal protrusion and cannot be slipped past its midpoint, which helps center it over the two tube ends.

When you solder copper (or any other metal) you depend on the solder wetting the copper. When this happens some of the copper is dissolved in the solder, and when the solder cools and hardens a permanent, atomic bond is formed. In a joint the solder lies between two pieces of copper, bonds to both, and so joins both. A good solder joint cannot be ripped apart. The solder will not separate from the copper: a layer of solder will always be left on the copper.

The key is wetting the copper with molten solder, and in order to do this, three conditions must be satisfied: the copper must be perfectly clean, covered with a layer of flux (that prevents the clean copper from oxidizing when heated), and hot enough to melt the solder completely. Fail in any of these three conditions and the solder will not wet the copper and bond to it. No joint or a very poor joint will be formed.

Material and equipment—To sweat tubing of any diameter, you need a propane torch. A soldering iron will not do: the largest (and they do come larger than 500 watts) will not furnish sufficient heat. Any standard American propane torch will do fine. You don't need a selection of nozzles. A nozzle 1¼ inch long and ½ inch in diameter is sufficient for most if not all sweating jobs. If the day is cold, if there is a wind and if the part to be heated is massive, no single nozzle will do—you need to use two torches simultaneously.

Use solid wire solder (50% tin/50% lead). The rosin-core and the acid-core wire solders should not be used because the rosin or acid flux tends to remain inside the long solder joints used for plumbing.

Use noncorrosive paste as a flux. There is a mixture on the market that combines lead (powder) and flux. It works very well, but is very expensive. There is no need for it.

STEPS IN SWEATING A JOINT

If you are going to work on a pipe that is part of a system, open the nearest faucet and leave it open until all the water drains out and until you have finished soldering. Should you heat a closed tube containing a little water you will soon build up a tremendous pressure, easily sufficient to burst the tube and send all nearby plumbers to the hospital. So always make certain the tube is empty and open at one end to release the steam that is generated.

If you are working on a faucet or a valve, or very close to a faucet or a valve, always disassemble it. Remove the stem and the washer. If you don't there is a good chance that you will melt the washer.

Do not place a wet rag on a nearby valve for the purpose of keeping its innards cool. The wet rag will keep the tube cool and will prevent a satisfactory joint from being formed.

Place the end of the tube inside the fitting or what have you. Check the fit. It must be loose (slip in and out without rubbing). If it is tight, solder will not enter. Use sandpaper to reduce its outside diameter, or get a new tube or fitting. Check to see whether or not the tube is oval. If it

36–39. Soldering a copper tube to a valve

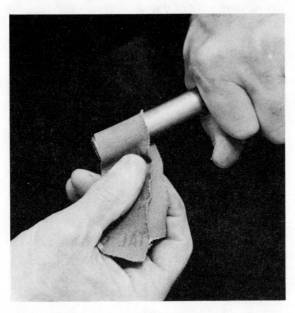

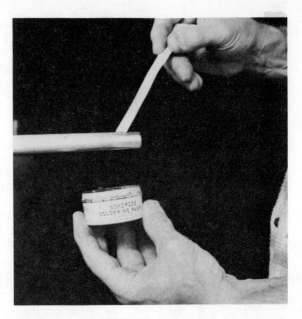

36. Clean the end of the tube with a piece of emery cloth, steel wool or sandpaper.

37. Apply a little soldering paste (flux) to the area to be soldered.

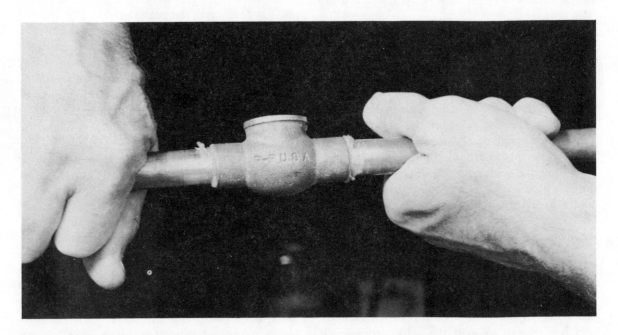

38. Take the valve apart. Slip it on the end of the tube. Apply heat to the valve and tube. Touch the tip of the solder wire to the joint. When the solder has been drawn inside, remove the solder and the heat. The joint is completed. Let it all cool down. Now push the other tube end into the valve body as shown.

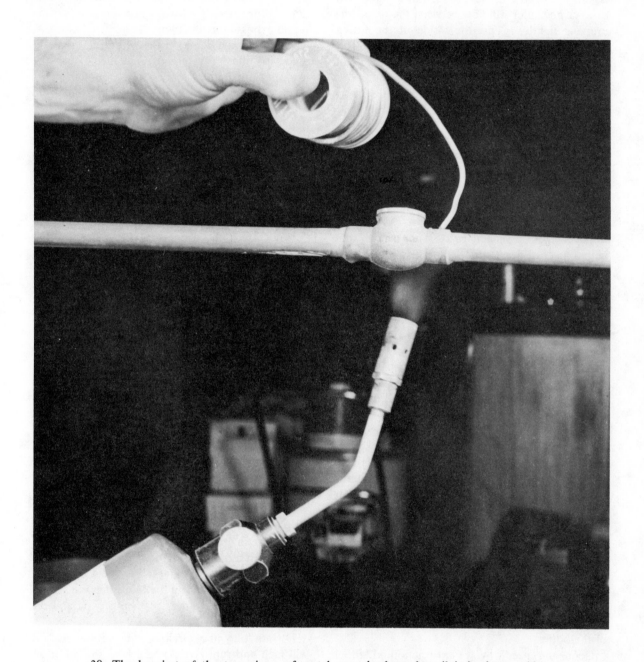

39. The heaviest of the two pieces of metal must be heated until it is above solder-melting temperature. Test by touching solder tip to metal, keeping both clear of the flame.

is and fits tightly in two places, those places will be dry: no solder will flow between the tube sides pressing tightly against the inside of the fitting. Should you go ahead with this joint, it will probably leak. Do not try to straighten the tube's end to cut a new end on the tube.

unless you have no other choice. It is much wiser

Rub the last inch or so of tubing bright with steel wool or sandpaper. Rub the inside of the fitting bright the same way. Do this even if both parts are factory bright. If the tube is old and corroded, be certain to remove all the corrosion; a few bright specks are not enough. Try not to touch the tube end or the inside of the fitting with either your hand or a glove.

Cover the brightened metal end of the tube with a layer of flux (soldering paste).

Place the tube end inside the fitting; give it half a turn to evenly distribute the flux.

Support the tube and fitting in one manner or another so that you do not need to touch either the tube or the fitting from here on out. Do not try to hold the tubing even a foot away from the point that is to be heated. Copper is an excellent conductor of heat. Should you hold the tube in your hand you will find that you will have to drop it long before the joint is finished.

Look behind the fitting that is to be heated. If the material there can be damaged by flame place a sheet of asbestos behind the fitting. If it is bare wood that will later be covered, it doesn't matter if it chars a bit as long as you soak it thoroughly with water as soon as you finish soldering.

Close the needle valve firmly on the nozzle portion of the propane torch. Be careful not to overtighten it, as doing so will permanently damage the valve seat and it will never close properly again.

Screw the nozzle portion of the torch onto the metal bottle. Doing so opens the valve at the top of the bottle.

Open the needle valve a fraction. Light a match and ignite the emerging gas. Next, open the valve further until you have a nice, steady flame. There will be a small roaring sound. If you open the valve too much, the rush of gas will blow the flame out. From this point on the torch is always held with the nozzle higher than the bottle. When the bottle is fresh and filled with

propane, tilting the tank above the nozzle causes liquid propane to be forced out the nozzle. The liquid extinguishes the flame.

Play the torch flame on the heaviest of the two pieces of metal that are to be joined. Keep the nozzle about 2 to 3 inches away from the part being heated. Try to keep the flame away from the joint opening. You don't want to heat the flux directly if you can help it. Keep the flame moving to heat the metal evenly.

Uncoil your wire solder. Move the flame off to one side. Touch the tip of the solder wire to the part you have been heating. If the solder melts, that part is sufficiently hot. If not, continue heating until it is.

Now direct the flame onto the smaller and lighter of the two pieces of metal. Keep playing the flame back and forth to heat this part evenly (but try to stay away from the open end of the joint). Move the flame to one side. Test the metal's temperature with the tip of the wire solder. If it melts, it is up to temperature.

Shift the flame back to the heavier of the two pieces of metal, taking care to keep the flame away from the joint opening. Now press the tip of the wire solder to the joint opening. If the joint is sufficiently hot, the solder will melt, and as it does it will be drawn into the joint by capillary action. You don't have to force it inside.

When the joint has drawn up all the solder it needs you will see a bright ring of solder all around the joint. Quit. Back the flame off and shut off the torch.

If solder drips out of the joint as you continue to feed it in, the joint is too hot. Back off the flame. Wait a moment and feed it more solder.

If the tip of the solder wire melts and drips down instead of entering the joint, the joint is too cool. Remove the solder, heat both parts and try again.

Do not play the torch on the wire solder: it will melt and run, giving you a false indication of joint temperature. Remember, the parts of the joint must be hot enough to melt the solder. Dripping hot solder on relatively cool copper will not form a bond.

Do not continue to heat the joint after it has been formed. It isn't necessary to pile solder up around a joint; all the strength lies between the faying (facing surfaces of the joint). Solder outside the joint has almost no strength. A properly made solder joint is stronger than the base metal (copper in this case) itself. If you pull hard enough on a sweated joint the copper tube will give before the joint is pulled apart.

Do not touch or move the joint until you see the solder chill. Its surface will lose its sheen and become slightly granular. Move the joint before this occurs and you will form a cold solder joint, which is very weak.

After the solder has chilled you can splash water on it to speed handling. Doing so will not harm the solder joint.

That is all there is to sweating a joint. After you have done it a few times the process will become almost automatic and you will be able to form them almost as quickly as you can heat the tubing.

When you have finished with the torch, unscrew the nozzle from the bottle. Doing so closes the valve in the top of the bottle and is a much safer way of storing the torch than depending on the needle valve to remain tightly closed. Store the bottle in a cool place.

TINNING

Tinning is the practice of coating metal (in our case, copper tubing and fittings) with a thin layer of solder. Tinning is used to ease and speed the sweating of joints that would otherwise be difficult or dangerous to make—for example, overhead joints. Tinning procedure is almost identical to soldering procedure.

The surfaces to be tinned are brightened with steel wool or sandpaper and covered with flux. Each part is then heated to the melting point of solder, whereupon the tip of the solder wire is pressed and rubbed against the fluxed portion of the metal until a thin layer of solder covers it. The heat is then removed and the parts are permitted to cool.

Assume that you are going to join a tube to an elbow. The tube's end would be tinned and the inner surface of the elbow would be tinned. Then both would be permitted to cool. The tube end would then be pressed a short distance into the opening in the elbow. Next, both the tube and the elbow would be heated. When the solder melted, the tube would be pushed inside the elbow and the heat would be removed. In this way the joint would be formed very quickly.

COMPRESSION JOINTS

All compression joints used in home plumbing systems are alike and are described below. They are also used for thin-wall brass pipe joints.

A compression joint consists of four parts: a fitting, which may be the enlarged end of a tube, the end of a faucet or valve or an individual fitting; a ring of rubber or metal; a compression nut or collar; and the tube (or pipe) itself.

Assembly—The clamping nut is slid up on the tube, and the ring is positioned. The end of the tube is slipped into the fitting (which may be an individual fitting or a part of something else). The ring is pushed as far as it will go into the fitting. The compression nut is pressed against the fitting until it engages the threads cut on the outside of the fitting. Then the nut is turned until it is snug. As the nut is turned it moves toward the fitting. The ring between the outside of the tube and the inside of the fitting is compressed and seals the joint between the tube and fitting.

Flexible-ring details—Flexible-ring compression joints are used mainly with large-diameter, thin-wall brass pipe for the drain connecting sinks, lavatories and tubs with galvanized or plastic drainpipes leading to the house drain or drain stack.

Flexible-ring compression joints can be opened and closed innumerable times because the flexible ring "gives" on being compressed and is itself not deformed, nor does it deform or damage the tube or pipe.

Little care is necessary in making and assembling a flexible-ring compression joint beyond making certain that the end portion of the tube is perfectly round and corrosion free. If the tube or pipe's end is rough or bent the joint may leak. Another minor requirement, easily met, is making certain that the tube enters the fitting for at least an inch or more. The joint will hold as long as there is sufficient tube to support the ring, but if the tube is very short a slight sideways push will open it up (see illustration 13).

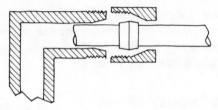

40. Interior of a compression joint utilizing a metal ring.

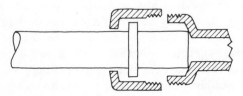

41. Interior of a compression joint utilizing a flexible ring. This type of joint is most often used with thin-wall pipe an inch or more in diameter.

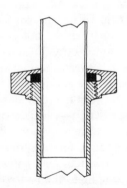

42. Slip-type flexible-ring compression joint. Inner tube can be slipped a distance into the outer tube, making it unnecessary to cut inner tube to exact length.

There is no difficulty in getting the correct flexible ring for the joint. The correct size—and not very many sizes are made—needs to be stretched just a fraction to make it fit. If it is loose it is wrong and if it is very tight, it too is incorrect. The incorrect nut will neither fit over the tube nor engage the fitting, so you cannot go wrong there.

Slip joints—Some flexible-ring compression joints are made to accept several inches of tub-ing. Some are not. Check by assembling the joint without tightening the compression nut. If the tube can be pushed into the fitting a good way, it is a slip joint. If not, it is not. The advantage of working with a slip joint is that you do not need to cut the tube to exact length, but can make it a little long. Then when you assemble the joint you adjust the tube's position within the slip joint to where you want it.

Metal-ring details—In assembly and in principle of operation the metal-ring joint is similar to the flexible-ring compression joint: the ring mounts on the tube and, together, tube and ring enter the fitting, where the ring is expanded by pressure exerted by the compression nut. However, the use of a metal ring in place of a flexible ring alters the assembly and utilization of the joint far more than one might imagine.

The metal ring fits the tube so closely that the passage of the ring will be stopped by corrosion, paint and even the smallest external burr. Therefore, if you saw the tubing, you must carefully file the burr off before assembling the joint. If you try to force the ring into place the joint may leak. If the tube end is oval for any reason, the same precaution applies: do not force the ring on. And, do not try to hammer the tube end into a round shape; you cannot do a perfect job with a hammer. If the tube end is less than perfect, the joint will leak. The practical solution is to cut the end of the tube off and work with a perfect section of tubing.

Should you need a bend in the tube, bend it before you assemble the joint, taking care to leave at least 2 inches of tubing perfectly straight. If you take the easy way, assembling the joint and then bending the tube using the joint as a fulcrum point, the joint will probably leak. The tubing must be straight from the inside of the joint to well past the compression nut.

After you have taken the described precautions, assemble the joint, making the nut just a little more than finger tight. If all is well the joint will hold water under high pressure. If it doesn't, take the joint apart and look for dirt and grit in the fitting, on the tube and on the ring. Reassemble; try pressure. If the joint still leaks, tighten the compression nut a little more. When leaking stops, stop increasing the pressure.

Anything more than light pressure deforms the metal ring and the tube it encloses. Once the ring is deformed it cannot be moved from the tube, and the tube cannot be rotated in the fitting (when loosened) without leaking when retightened. The degree of deformation naturally depends on how tight you make the compression nut. If you have no choice but to apply lots of pressure, remember, should you need to take the fitting apart you must put it together exactly as it was before; even then it may leak.

In some instances you will assemble a metal-ring compression joint and it will leak although all necessary care has been taken. This is probably due to a poorly made ring or fitting. Either replace the tube and fitting and start all over again or see Chapter Ten, Fixing Leaking Pipes and Joints.

43. Note how closely the metal ring fits the tube. This is what makes it possible for a metal-ring compression joint to hold tightly with only a little tightening of the compression nut.

In recent years a plastic ring has been introduced replacing the metal ring used in metal-ring joints. When possible, always opt for the plastic ring (usually black): it is much better than the metal for the purpose.

FLARE JOINTS

A flare joint consists of three parts: flange nut, fitting and tube. Flare joints are like compression-ring joints in that the flared end of the tube is compressed against a small cone inside the fitting. However, a flare joint doesn't use a ring of any kind to seal the joint.

Flare joints are used when and where it is inconvenient to solder, where solder will eventually lead to galvanic destruction (in moist soil) and when the joint may need to be taken apart at some future time. The only reasons that flare joints aren't usually used in place of metal compression-ring joints is that flare fittings cost much more and take time to make.

Making a flare joint—A flaring tool is required, but it isn't very expensive. Start by selecting a fitting sized to the tube. Cut the tube, if necessary, just as squarely as you can. Remove the external burrs, if any, and then remove the internal burrs taking great care to remove the burr and nothing more. You don't want to bevel the inside edge of the tube.

Slide the flange up and over the end of the tubing. The tubing end is placed in the vise portion of the flaring tool and clamped firmly in place. The tip of the tool's ram is cleaned and given a drop of oil or a smear of wax. Then the ram is screwed down into the tube's end, flaring it (forming a cone shape). The tool is removed. The flare is pressed against the cone inside the fitting. The flange nut is slid down and rotated against the fitting. Doing so presses the flare against the cone, effecting a watertight seal.

Flare details—The joint is formed between the flared end of the tube and the inner metal cone. The contact area is very small, so you have to be careful that the flare on the tube end is clean when you form it, that it isn't too small (with insufficient metal making contact), and that it isn't too large. In the latter case you won't be able to get the flange nut in place.

44–47. **Making a flare joint**

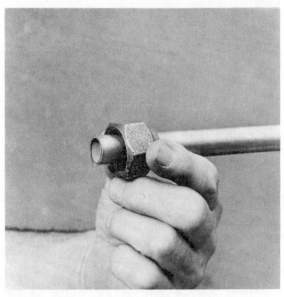

44. First step in making a flare joint is to cut the tube, ream its inside free of internal burrs, then remove the external burrs. Next, slide the flange nut up and over the tubing. Note where its threads go.

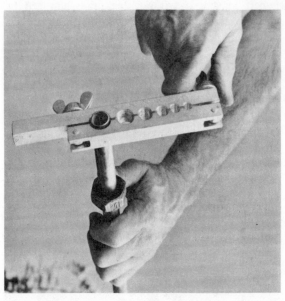

45. Place the end of the tube in a flaring tool and clamp it firmly in place.

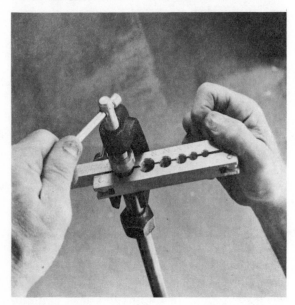

46. Force the tool's ram down into the end of the tube as far as it will go.

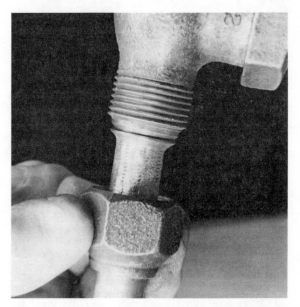

47. Remove the tube from the tool. Press the flared end against the fitting and screw the flange nut into place.

Flare size is dependent on the position of the tube in the vise. In some tool design the tube end must be flush with the face of the vise. In others the tube must extend a fraction of an inch beyond the vise face. If your gadget doesn't carry instructions, run up a few test flares and see which is correct by comparing the flare area to the cone area.

After you have made a half dozen flare joints or so, their making will become automatic; no problem at all. They can be taken apart and reassembled without trouble as many times as you wish.

SWAGGED JOINTS

A swagged joint consists of an expanded copper tube end into which the end of an unchanged-diameter tube is placed. The two tube ends are then sweated together.

The swagged joint eliminates the tube coupling. However, you do need a swagging tool, which, while relatively inexpensive, still costs more than a dozen copper tube couplings. Unless you need a great many such joints, then, the coupling is the best deal.

To form a swag, the tube end is placed in the

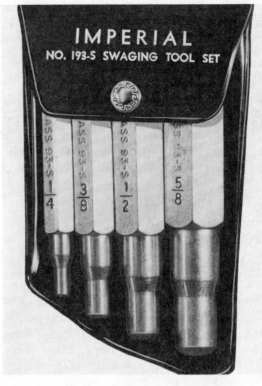

48. Typical set of swagging tools. Each tube size requires its own tool. *Courtesy of Imperial Tool Corporation.*

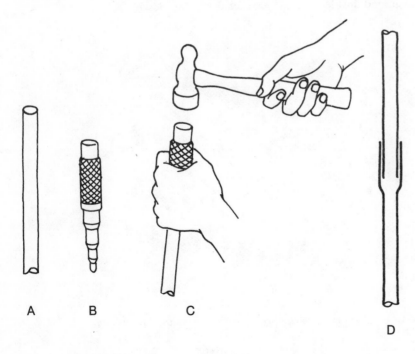

49. To join two tube ends by swagging, take one tube end (A) and hammer the end of the swagging tool (B) into it as shown (C). The second tube can now be slipped a distance into the end of the first tube and soldered in place (D).

vise portion of the tool. Depending on tool design, you either drive a shaped bar down into the tube end with a hammer or you force it down with a ram arrangement similar to that used with a flaring tool. In fact, some flaring tools have adapters which enable them to swag.

When the tube's end is swagged, it is removed from the vise. The inside surface of the expanded end is cleaned with steel wool or sandpaper. The same is done with the end of the tube that is to be inserted. Then the mating surfaces are fluxed and the joint is sweated.

BENDING TUBING

One of two tools is necessary for bending tubing; a wire-spring bender or a fulcrum bender (often called a hickey by electricians). The wire-spring tool can only be used on soft copper tubing; however, rigid copper tubing becomes annealed (soft) after it has been heated and allowed to cool. The fulcrum bender can be used with either soft or rigid copper tubing.

Wire-spring benders are manufactured in a number of diameters to fit various diameter tubes because the fit between the spring and the tubing must be close. The spring prevents the tube from kinking (flattening). The tube is slipped inside

the spring and hand pressure is applied until the desired bend is secured. That is all there is to it.

To use the fulcrum bender, the tube is slipped in the groove in the quadrant and beneath the hook at the end of one of the tool's handles. Then the two handles are brought together and

50. Using a tubing bender. The tube is slipped into the bender, then the bending action is done by hand. Tool prevents tube from flattening, but you must use a tool that fits the tubing closely.

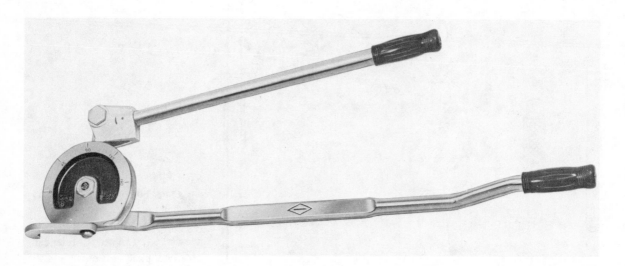

51. Typical fulcrum tube or pipe bender. *Courtesy of Imperial Eastman Corporation.*

you have a 90-degree bend; less if you just bring the handles partially together.

You can also bend copper tubing with the aid of a vise, but it makes for a very rough-appearing arc and there is a good chance of kinking the tube, which renders it almost useless. However, the vise is something to keep in mind for an emergency.

The tube is slipped between the jaws of the vise, which are not tightened. The tubing is bent just a few degrees, then moved a fraction of an inch out of the vise. The tubing is again bent a few degrees, with the second bend at a point different from the first. This is repeated—bend a little, move a little—again and again until the desired curve is secured (see illustration 31). With care you can avoid flattening the tube, but it is not easy.

CHAPTER SIX

Working With Plastic Pipe

Wherever and whenever local plumbing codes permit, plastic pipe is *the* pipe to run for a good number of reasons.

Plastic pipe is much lighter than equal-size metal pipe. The weight difference between metal pipe and plastic pipe can be as much as 20 to 1. Since much of plumbing consists of simply moving pipe from here to there and back again, plastic pipe is far easier to work with, especially in the larger diameters.

Plastic pipe is more easily joined than any of the other types of pipe. A swab of solvent cement (a kind of glue) makes an almost instant, permanent, watertight and pressure-tight joint. The cemented joint is just as permanent and dependable as solder, but much faster and safer. No flame is used and there is no waiting for the parts to come up to solder-melting temperature.

Screw-thread joints can be made just as rapidly, but only after the pipe has been cut to length and threaded, both of which tasks require considerable time, effort and expensive tools.

Plastic pipe is easily cut to size with inexpensive tools. Rigid plastic pipe can be cut with a hacksaw, a carpenter's saw or a standard pipe cutter. Flexible pipe can be cut with a saw or a sharp knife.

Plastic pipe is electrically inert and does not form a galvanic couple with any known substance. Therefore plastic pipe will not corrode when buried in wet earth alongside or even touching metal of any kind.

Plastic is inert. It disintegrates very, very slowly. No one knows just how long it will take a plastic pipe to fall apart, but certainly it should last as long as metal above or below the ground when exposed to normal atmospheric moisture.

Plastic is a thermal insulator. Plastic pipes carrying cold water through a warm room do not sweat as much as metal pipes do. Hot water flowing through plastic pipe does not lose as much heat to surrounding air as hot water flowing through metal pipe. There is far less need to insulate the pipes when they are of plastic, because it is an insulator.

Like most insulators, plastic pipe is a poor conductor of sound (as compared to metal). The opening and closing of valves and faucets in one portion of a building will not be as strongly carried to another portion of the building if plastic piping has been used in place of metal.

Best of all, plastic pipe is cheaper.

The foregoing are the major advantages of plastic pipe over metal pipe. As is true of most things in life, there are some disadvantages to plastic pipe. The single greatest disadvantage is the reluctance of many local plumbing authorities to permit its use, because union leaders oppose the use of plastic pipe, which greatly reduces the hours of labor required to do a job. (If plastic pipe is not permitted in your town and you install it, the authorities can force you to remove it. The alternative may be a daily fine.)

Additional but far less troublesome disadvantages include plastic's relatively large coefficient of expansion. Roughly, plastic pipe expands 1 per cent lengthwise for every 100° temperature rise. In more specific terms, a 10-foot

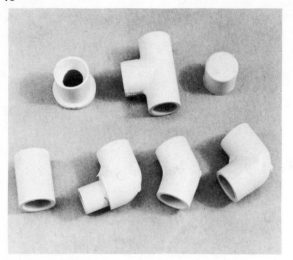

52. A few of the many plastic fittings used for running plastic pipe. The coupling at top left is a reducing bushing.

length of plastic pipe becomes 1 inch longer when it is heated 100°. This would be the approximate temperature change in the piping connected to a hot-water-heating boiler between the boiler's turned-off and turned-on condition.

Plastic pipe also softens as it becomes warmer. If plastic pipe is heated beyond its rated capacity and held there for a period of time, the pipe will sag and will remain looking like a tired snake when it cools.

As can be expected, when plastic pipe softens with heat its ability to withstand internal pressure decreases. To put it technically, its pressure rating decreases with increasing temperature.

Commercial grade plastic pipe cannot withstand temperatures as high as metal pipes can. The long-term temperature rating of CPVC plastic pipe under a pressure of 100 psi (pounds per square inch) is only 180° F. However, the pipe can withstand slightly higher temperatures for short periods of time if the internal pressure doesn't go up simultaneously. Other types of plastic pipe can also withstand continuous exposure to 180° F. water, but at much lower pressure.

Plastic pipe, like copper, must be protected from stray nails and foot traffic. While plastic is physically stronger than copper (at least in the thicknesses used) it cannot be indiscriminately exposed, as can galvanized and cast-iron pipe.

Practical considerations—Plastic's high coefficient of expansion is easily accommodated by using loose-fitting clamps to hold the pipe. Plastic pipe's tendency to sag can be offset to some extent by using lots of supports. Plastic pipe's thermal limitations are circumvented by using several feet of metal piping to make the final connection between the plastic and the hot-water coil in the furnace or the hot-water-heating tank. Hot water supplied to the various fixtures and washing machines should never be as high as 180° F., so no difficulty is encountered there.

PLASTIC PIPE FORMULATIONS

The formulations or types of plastic pipe most often used in the home are ABS, PVC, CPVC, and PB.

ABS (acrylonitrile-butadiene styrene) is generally black or gray in color. It is used primarily for drains, wastes and vents. It is rigid.

PVC (Polyvinyl Chloride) is generally cream colored. It is rigid and generally used for cold water lines inside and outside the house.

CPVC (Chlorinated Polyvinyle Chloride) is generally cream colored. It is rigid and most often used for hot- and cold-water lines inside the house.

PB (polybutylene) is generally black or dark gray and is most often used for water service, water mains, gas service and irrigation systems. It is flexible.

WEIGHT OF PLASTIC AND METAL PIPE IN POUNDS PER HUNDRED FEET

Size	PVC	Copper DWV	Galvanized iron	Cast iron
1¼	42	65	227	—
1½	51	81	271	—
2	68	107	365	400
3	141	169	750	600
4	201	287	1,070	800

MEASURING PLASTIC PIPE

Like all other pipe and tubing, the various sizes of plastic pipe are based on their internal diameters. The size assigned is whatever fraction or whole number is closest to the actual dimension. For example, the internal diameter of 1½-inch PVC pipe is actually 1⁹⁄₁₆ inch—¹⁄₁₆ inch more than its nominal size. Its outside diameter is 1⅞ inches.

To find the length of pipe necessary to connect two fittings or a fitting and a valve a particular distance apart, measurements are made exactly as they are made for threaded pipe (described in Chapter Four under Measuring pipe).

Repeating the instructions briefly: the distance the pipe end will enter the fitting is found by actually slipping the pipe into the fitting and/or valve. Then the two fittings or the fitting and the valve are positioned the exact, desired distance apart. Next the distance between the facing edges of the fittings is measured and this dimension is added to the distances the pipe's ends enter the fittings. In this way the overall pipe length necessary is determined.

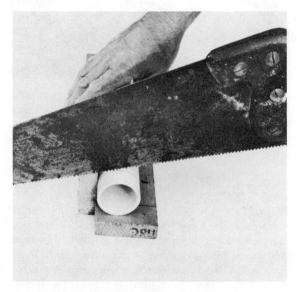

53. Large-diameter plastic pipe is easily cut with an old saw. A simple jig helps keep the cut square.

CUTTING PLASTIC PIPE

If you are using a saw it is best to work with a miter box or similar guide to make certain that your cut will be fairly square. If you are working on flexible pipe and using a sharp knife, wrap some white adhesive tape around the pipe to guide you. If you are using a pipe cutter, the cutter itself ensures a square end.

After cutting, the internal burr must be removed, and if there is any external unevenness, that too should be removed. All this is easily done with a pocket knife.

JOINING PLASTIC PIPE

Solvent welding—This is the term applied to the technique of cementing plastic pipe to plastic fittings. The term "welding" is fairly accurate, as properly cemented plastic-to-plastic joints are permanent and solid.

Only a few simple steps are necessary to make a good solvent-welded joint, but like those necessary for soldering, each step is important and overlooking or slighting one can lead to a leaking joint. Unlike solder joints, however, solvent-welded joints cannot be taken apart and remade. The leaking joint has to be cut out of the line and replaced. (See Chapter Ten, Fixing Leaking Pipes and Joints.)

The required steps in making a solvent-welded plastic joint are as follows:

Measure and cut the plastic pipe to length. Remove internal and external burrs.

Try the pipe in the fitting. The fit between the two must be very close, so close you can "feel" the pipe entering the fitting, but not so close the pipe has to be forced. The hole must be perfectly circular. Test by rotating the fitting on the pipe end. If the fit is not close, or if the fitting binds when you rotate it on the pipe, discard that fitting and try another.

Position the fitting correctly on the pipe. Use a pencil to mark the alignment on the fitting and the pipe.

Take fine sandpaper or emery cloth and remove the shine from the pipe's end for the distance the pipe will enter the fitting. If the fitting is

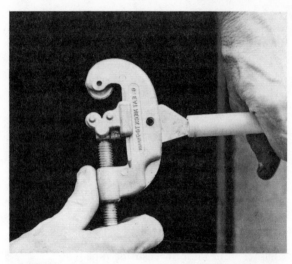

54. The plastic pipe is cut, with either a saw or a tube cutter, as shown.

55. The inner burr is reamed clean. External protrusions, if present, are removed with a small knife.

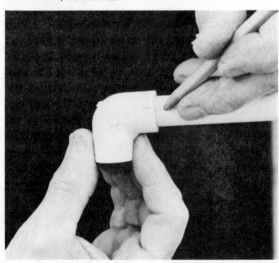

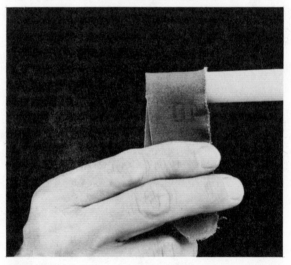

56. The fitting is placed on the tube end. It must make an interference fit—meaning that you can feel it go on. Mark both the fitting and the pipe end so that you will know the fitting's position.

57. Remove the shine from the end of the pipe with some fine emery cloth. Do the same with the fitting socket.

58. Apply a very thin coat of cement to the end of the pipe. Do the same with the fitting. Give them a minute to dry. Follow with a second, thick but even coat on the pipe end alone. Position the fitting on the pipe end. Give the fitting a quarter turn or so to make certain that the cement is evenly distributed. Bring the two pencil marks into alignment. Let the cement harden. The joint is completed.

large enough, do the same for the inside of the fitting. Just a few strokes with the sandpaper does it. Do not grind away and reduce the pipe's external diameter.

Blow the dust away. Take care to keep hands and water away from the pipe end and fitting interior after sandpapering.

Quickly apply a very thin coating of cement to the end of the pipe for the distance it will enter the fitting. Let a minute or so pass. This thin coat acts to soften the plastic. Follow with a second, thick but even coat of cement.

Insert the pipe into the fitting. Give the fitting a half turn or so to distribute the cement. Align the pencil mark on the fitting with the pencil mark on the pipe.

Let the joint rest for at least 30 seconds in warm weather, longer when it's cold. Now you can continue to "make up" the plastic pipe run, but do not fill the line with water under pressure for several hours. Most plumbers let the cement harden overnight before applying pressure to a newly made plastic joint.

Cementing tips—Work fast when you apply the cement. It dries very quickly. Use a brush just as wide as the width of the cement band you require. Be certain to use a hog-bristle brush. Plastic brushes usually are dissolved by the cement.

Always make the second coat a little thicker than necessary. When a bead of cement is forced out of the joint as you make it, you know there is sufficient cement.

Always keep your cement can tightly closed when not using it. Should the cement get thick and ropy, discard it.

Make certain you use the correct cement. Use ABS cement for ABS pipe, PVC cement for PVC pipe and CPVC cement for CPVC pipe.

Keep your cement warm. Should you be forced to work in cold weather, store the cement in a warm place; an insulated box kept warm by a hot dry brick can be used.

Keep the joints and cement dry. Even a few drops of perspiration in the cement can interfere with joining.

A few plumbers apply a coating of solvent to the surface of the pipe ends in place of sanding and in place of the first thin coating of cement.

Some do this in addition to sanding the shine off the plastic. They apply the solvent with a brush or rag. Whether or not this ensures a better joint is doubtful, as most plumbers do not trouble using solvent in addition to cement.

Threaded joints—Plastic pipe can be threaded just like any other pipe. However, it isn't done very often because threading tends to weaken the pipe. When plastic pipe is threaded it is usually given a male thread so that the plastic pipe is reinforced by the surrounding metal pipe.

Clamped joints—Flexible plastic pipe is usually joined by clamps, though it can also be joined by thermal welding. The clamped joint is made by forcing the flexible plastic pipe over a stainless steel or plastic insert, which is then held in place by clamps similar to those used on automotive water hoses. The system is fast, dependable, relatively inexpensive and easily opened. Usually a screwdriver or a socket wrench is the only tool needed.

Compression joints—Rigid plastic pipe is often connected to various fittings by means of a flexible-ring compression joint (fitting). These joints are very similar to those used with metal pipe. The joint or fitting itself consists of a flexible plastic ring, a support and a compression nut. The fitting may be a portion of a plastic trap or it may be an individual fitting which is cemented

59. Flexible plastic pipe can be joined by means of a stainless steel coupling and a pair of clamps.

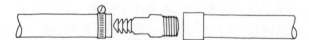

60. Flexible plastic pipe can be joined to threaded pipe by means of an adapter, a clamp and a coupling that goes on the end of the male end of the pipe. If you heat the flexible tubing with hot water prior to making the joint you will find it easier to get the plastic up and onto the adapter.

61. Using a plastic flexible compression-joint fitting to connect a plastic drainpipe to a plated thin-wall brass tail-pipe (connects to a sink or a lavatory drain).

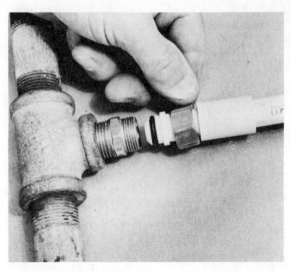

62. Using an O-ring adapter to connect a plastic pipe to a threaded Tee. The use of an O-ring adapter here permits the plastic pipe to be disconnected from the threaded fitting. An ordinary plastic-to-threaded-pipe adapter cannot be disassembled once the plastic is cemented in place.

onto the pipe or trap end. Like the flexible-ring compression fittings used with metal pipe, the plastic fittings can also be adjusted and taken apart as many times as required without any problem.

Connecting to other pipes—To connect a plastic pipe to a galvanized-iron pipe or to copper tubing an adapter (transition) fitting is used. One end of the adapter accepts the end of the plastic pipe, which is cemented in place. The other end of the adapter may be male-threaded to screw into a coupling or any other female-threaded opening, or the second end of the adapter may have a molded-in-place collar and an O-ring. The O-ring end fits into a smooth-bored hole such as is used with a metal-ring compression fitting. For example, an adapter would be used to connect a plastic pipe directly to a shutoff valve.

Connecting an O-ring adapter—The adapter fitting is "sized" by the size of the plastic pipe it is designed to accept. The other end of the adapter is designed to fit the same size opening. For example, to connect ½-inch plastic pipe to a

½-inch shutoff valve, you would use a ½-inch adapter.

To make the connection, slip the compression nut over the end of the plastic pipe. Cement the fitting onto the end of the pipe. Place a drop of Vaseline or other mineral oil or grease on the O-ring. Insert the O-ring end of the adapter into the metal valve or what have you. Bring the compression nut up and make it snug. Do not overtighten as you can easily damage the adapter. If the joint leaks there is dirt between the O-ring and the metal, or the metal opening isn't perfectly round. Try another valve in its place.

RUNNING PLASTIC PIPE

There is nothing special about running (installing) plastic pipe. You follow the same general procedure you would use with galvanized.

Fittings—One plastic fitting has already been discussed: the adapter. The rest are almost identical to their galvanized-iron counterparts. There are plastic elbows, couplings, unions, Tees, Wyes, reducers, bushings, caps and plugs. However, there still are no plastic valves or faucets. Therefore, to introduce a valve into a plastic pipeline you have to use an adapter fitting to either side of the metal valve.

There are, however, plastic traps and associated plastic, flexible compression-ring fittings designed to accept thin-wall brass pipe such as is commonly used with fixture drains. Thus, by selecting the correct fitting you can go from a plastic trap to a thin-wall brass tail-pipe (sink or lavatory drainpipe). There are also flexible compression-ring fittings that will enable you to go from a plastic drainpipe to a brass trap with no problem.

Bending plastic pipe—Flexible plastic pipe can be bent without difficulty to a fairly small radius. The rigid plastic pipe can be bent a little more than galvanized, which is almost not at all. However, if necessary, rigid plastic pipe can be bent to a small radius with the aid of heat. It isn't easy, but it can be done. The pipe is tightly filled with sand to prevent it from kinking. Then the area to be bent is carefully heated with a propane torch. When hot, the pipe is bent as desired. Upon cooling the plastic retains its bend.

Supporting plastic pipe—Clips are manufactured expressly for supporting plastic pipe. They differ from clips or straps designed for metal pipes in that the plastic-holding straps do not bind the pipe in place but hold it loosely, permitting the plastic to expand and contract at will.

Plastic pipe carrying hot water and running horizontally should be supported every 3 feet or more. The same pipe carrying cold water needs only to be supported every 6 feet or so. Hot- or cold-water-carrying plastic pipes that run vertically can be supported every 5 feet or so.

CHAPTER SEVEN

Curing Toilet Troubles

The modern flush toilet, invented in England in the middle of the nineteenth century, looks like Rube Goldberg's best work: a contraption consisting of a confusion of floats, levers, valves and hissing, rushing water. Actually, however, it is an economically simplified version of a rather sophisticated piece of machinery. It is for this reason that flush toilet mechanisms are difficult to adjust and repair without first stopping to consider their operation.

HOW IT WORKS

The most common type of toilet flush-tank mechanism consists of a **ball cock valve, stopper, control lever** and attached **wire rods, overflow pipe** and **refill pipe.**

The ball cock valve is connected to the water supply pipe. This valve controls the flow of water into the tank. The valve itself is controlled by a float, which is attached to the valve by a rod that can be bent for adjustment. When the tank is emptied and the float drops, the ball cock valve is opened and water flows into the tank through the ball cock valve and refill tube. As the tank fills with water the float rises. When it reaches a preset point, the ball cock valve is closed and water ceases to enter the tank.

When the control lever is pushed an attached arm and wire lift the stopper (a rubber ball). From this moment on the stopper rises through the tank full of water and rides on the surface. This is why you do not need to hold the control lever down all the while the tank is emptying into the bowl.

The weight and speed of the water flowing down into the bowl through the large opening exposed by the rising stopper is such that its momentum carries it through the trap. Thus the soil in the bowl and a goodly portion of the water is carried up and over the lip of the trap and down into the soil pipe. When the tank is empty and no more water flows into the bowl, the bowl is clear of soil and much of its normal complement of water.

At this moment, the stopper which has been descending on the surface of the water reseats itself over the opening leading to the tank. This opening serves as a valve seat.

At the same time, the float attached to the ball cock valve has also descended with the lowering water level. As the float does so, the ball cock valve is opened and water rushes into the tank. However, since the stopper is now sitting in its seat (the hole leading to the bowl), water can no longer run out, so the tank is gradually filled up.

At the same time that water emerges from the ball cock valve, water also emerges in a fine stream from the refill tube, which is pointing down into the overflow pipe. The small stream of water replaces the water lost in the toilet bowl and ensures that there is always water in the bowl and in the communicating toilet trap.

When the float rises to its preset position, the ball cock valve is shut off once again. Water stops flowing into the tank and into the bowl.

63–66. **How a toilet works**

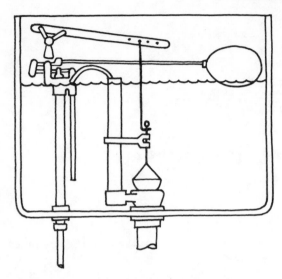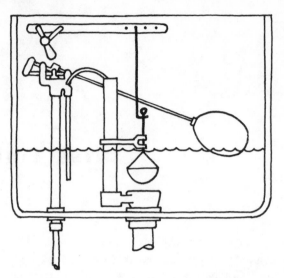

63. The mechanism is in its rest position. The tank is full. The ball floats high, and no water is entering or leaving the tank.

64. The control handle is operated. The stopper ball is lifted clear of its seat, whereupon it floats on the surface of the water. The water in the tank rushes into the bowl. The water level drops, the float descends, and as it does so it opens the ball cock valve. Water enters the tank from the water supply line, but not fast enough to keep the float from descending.

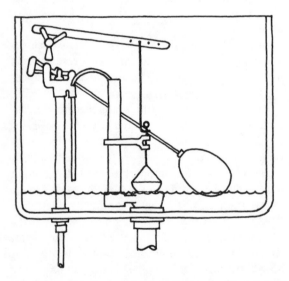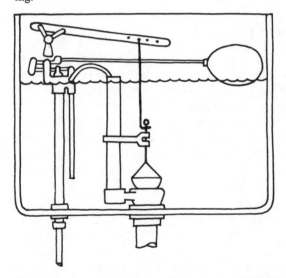

65. The water has run out of the tank. The stopper ball rests on its seat. Water enters the tank, but since the stopper ball's bottom is open, it cannot float from this position. Water continues to enter the tank; some of the water passing through the refill tube enters the overflow tube and refills the bowl.

66. When the water in the tank has risen so high that the upward pressure of the float closes the ball cock valve, no more water enters the tank and the mechanism rests until the control handle is again operated.

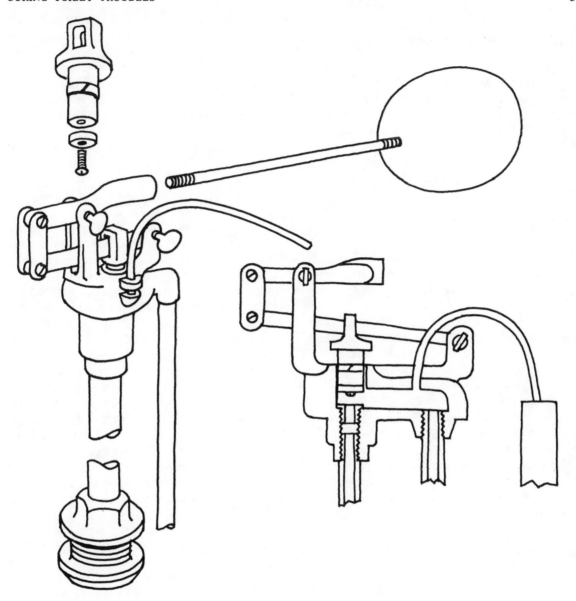

67. Major parts of the "common" type of toilet-tank mechanism.

The mechanism then rests until it is called upon to perform its duty once again. Should the ball cock valve fail to shut off completely, water entering the already filled tank will not run out onto the floor but will run into the overflow pipe and safely into the toilet bowl.

Newer flush-tank toilet mechanisms depend on a diaphragm valve for incoming water control. The sequence of operation is identical to that just described. The only difference lies in the valve mechanism. The same arrangement of float control is used.

TOILET MECHANISM PROBLEMS AND CURES

Water won't stop running (Ball cock valve)—Remove tank cover. Check on water level. If water level is above the top end of the overflow pipe, the ball cock valve is not working for any of several reasons: the float may be set too high, in which case the float isn't raised high enough by the water level to shut the ball cock valve off. Check this by hand-lifting the float. If water inflow stops, use both hands to bend the float rod slightly downward. Flush the toilet and observe the results.

Float may be inoperative due to leak—it doesn't float. Unscrew float, shake to detect presence of water. If leaky, replace. Try tank operation with new float.

Float rod may have turned, permitting the float to move upward. Try rod to see that it is tightly screwed into the valve arm. If it isn't, correct.

Float arm linkage may be defective. Empty tank by lifting control lever. Raise and lower float and see if mechanism works. If linkage is defective you can sometimes make a temporary repair using a nail or a bolt as an axle (to hold moving parts together).

Ball cock valve may be defective. If lifting ball by hand with moderate pressure doesn't stop inflow of water, it is the valve itself that needs repairs.

TO REPAIR BALL COCK VALVE

Close stop valve beneath toilet bowl. If there is no valve find nearest valve in line and close it, or go to main valve in basement and close it. If you do this, drain the cold-water line by opening a top-floor faucet and a basement faucet.

Disassemble the ball cock valve pantograph mechanism (the levers atop the valve and attached to the float by the rod). It's best to do this when you know that a plumbing supply store will be open, as old brass and copper parts will sometimes crumble under your hand.

Lift the ball cock plunger up and out. Examine the washer, a circular disk of rubber pressed into its bottom. If it is defective replace it. It is merely pressed into place. Use a flashlight and look down the hole whence the plunger emerged. If the seat there shows cracks or a very rough surface, the seat is defective and must be replaced. Use a valve-seat removal tool and a wrench to remove it. If you can't unscrew it, the valve seat is either rusted fast or is a part of the valve and cannot be removed. In either case the only solution is to replace the entire ball cock valve mechanism.

TO REPLACE BALL COCK VALVE

Make certain that the flow of water into the tank has stopped and that your shutoff valve truly holds. Some of them do not. In such cases go to the main valve or an intermediate valve. If you don't you will have water dripping over the floor as you work. Drain the tank by working the control lever. Use a sponge or rag to remove the remaining water (it doesn't need to be completely dry though).

Use a crescent wrench to loosen the locknut on the metal-ring compression joint atop the stop valve below the tank. If there is no valve there, disconnect the feed pipe. Next, disconnect the feed pipe where it connects to the underside of the ball cock valve. This is another locknut. Remove the feed pipe, noting carefully just how it was installed. (If it is ⅜-inch tubing with a metal-ring compression joint at the bottom you cannot turn it around or bend it where it enters the lower joint or it will leak.) Next, loosen and remove the large nut on the underside of the ball cock assembly, the nut next to the tank. Now lift the ball cock assembly up and out.

Install the new assembly by simply inserting it in the hole occupied by the previous unit. Make certain that the large rubber washer or grommet is in place at the bottom. Now run the large nut back up the underside of the assembly. Make it snug, not tight. Hold the assembly with one hand to keep it from turning as you work on the nut. Remove the float and rod from the old valve and attach it to the new mechanism. Raise and lower the float by hand to make certain it doesn't hang up on any of the other parts in the tank. Now carefully attach the refill tube. It screws into a hole at the top of the ball cock valve. When it is

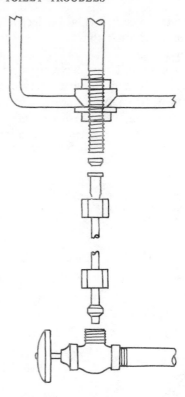

68. To disconnect the ball cock valve in a toilet tank, the feed pipe is first disconnected, then the locknut at the bottom of the tank is backed off. Now the ball cock valve can be lifted up and out. This large locknut is never made more than snug during reassembly; it is not necessary to make it tighter.

Water won't stop running, water below overflow-pipe top—The cause is at the stopper. Its bottom or seat may be corroded or covered with growth which prevents a perfect seat. Water seeps through; the tank is never filled; the float cannot shut off the water. Or, the ball may be punctured, permitting water to leak through. The lift wire may be bent or hung up in the guide so that the ball does not drop all the way down and fully close the opening.

To clean the seat, use steel wool and rub the opening down. The same can be done for the bottom of the stopper ball if it is merely dirty. If it is partially rotted, if it is leaking or if the lift wire is bent, replace both. The wire unscrews from top of the stopper.

Replace with a duplicate stopper or use a new flap-type stopper. To install the stopper, empty the tank; remove the old guide, the stopper and the wire. Clamp arms of the new flap stopper lightly to the overflow pipe. Rotate the new stopper until the flap lines up with the opening and is a fraction higher. Tighten the arm bolt. Hook the chain onto the control lever. There should be a little slack when the lever is in the rest position and the flap is down. Try the lever several times with water in the tank to make certain that it operates properly.

tightly screwed in place, bend it over so that it points down the overflow pipe.

Reassemble the feed pipe. Try not to bend it as you return it to its original position. Tighten the upper compression nut, again holding the ball cock valve to make certain it doesn't turn. Now replace and tighten the lower compression nut.

Open the water supply line and watch the ball cock valve mechanism operate. The float should shut off the water supply when the water in the tank is up to the mark (if provided by the manufacturer) or about 2 inches from the top. If the valve is shut off early, bend the float rod upward. If the valve shuts off late (water too high), bend the float rod downward.

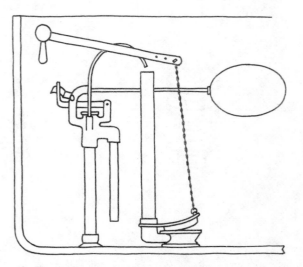

69. Depending on its design, some flap stoppers can be installed without moving the valve-seat assembly; other designs cannot.

Water won't stop running (Diaphragm valve) —Remove the tank cover. Check the water level. If the level is above the overflow pipe, check the float and the float arm as discussed. Lift the arm and apply light pressure upward. If this doesn't stop the flow of water, the valve is defective. The common defect is that the diaphragm is ruptured. Purchase replacement kit for particular valve.

Shut off water to tank. Replace defective part(s). Tank usually doesn't have to be completely drained to do this.

If no repair kit is available, drain the tank completely, as discussed, and replace the diaphragm valve assembly, following the steps outlined for ball cock valve replacement, with one precaution: check the connection requirement on

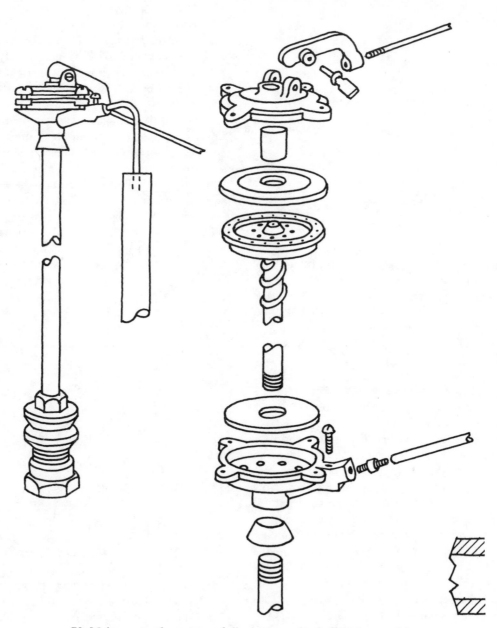

70. Major parts of one type of diaphragm-valve ball cock assembly.

71–72. American Standard diaphragm-valve ball cock assembly

71. The disassembled valve. The diaphragm is still within the valve body. The small round rod in the center of the valve's top (between fingers) moves up and down in response to the float. When this rod sticks the toilet sometimes "sings." Try loosening the rod with lubricant.

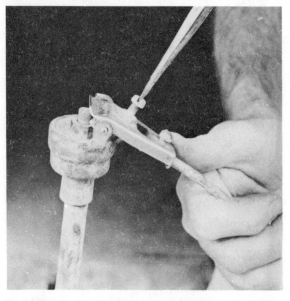

72. Assembled valve. Screw is adjusted to limit float drop and so limit rate of water flow into tank. When water pressure is low, rate of flow will be low no matter how this screw is positioned.

the feed pipe end of the new mechanism against the existing feed pipe. New toilet valves, ball cocks and diaphragms are all designed to accept a Speedee toilet fitting, which is a ⅜-inch tube with a preformed end. The end looks like a tiny trumpet mouthpiece and carries a plastic insert, a sort of dished washer. Very old toilets have a different fitting at this point, so make certain either that the water feed-pipe connection will fit the new valve or that you have a replacement fitting before you take your toilet apart.

When replacing a toilet valve you can install either a ball cock valve or the newer diaphragm valve. The hole in the bottom of the tank will accept either valve.

Water won't stop running (Diaphragm valve) —Water level is below top of overflow tube. Check stopper valve and its seat, check lift rod and guide as suggested for ball cock valves. Stopper operation is identical.

Control lever sticks in down position— Remove the tank cover. Operate the lever several times. If the arm hangs up on other mechanisms, bend slightly as necessary. If the arm is free of obstructions, the trouble is friction due to corrosion where the arm passes through the tank. There is no easy cure; it is best to replace the part. Remove the large nut holding the control lever to tank. Remove the lift wire or chain. Replace it with a new lever. Make certain that the lever doesn't touch anything. Replace the wire or chain, using the hole in the arm directly above the stopper, or close to it.

Control lever must be held down to flush—The stopper ball is filled with water and will not float. Replace it.

Stopper lift wire is bent and cannot be completely lifted—The stopper lift wire guide is set too low on the overflow pipe. The stopper cannot be lifted high enough to enable it to float.

Water squirts from beneath toilet tank cover— Should you lift the tank cover to investigate this phenomenon you may get a face full of clean water. The trouble is due to a worn ring on the plunger. This is a split leather or plastic ring (ring gasket) that fits in a circumferential groove in the body of the plunger. Like a car's piston ring, this ring seals the joint between the moving plunger and the walls of its cylinder. If the ring is worn, when the toilet is flushed the plunger is raised, admitting water to the cylinder, and some of the water gets past the ring and shoots upward against the tank cover.

Sometimes you can purchase a replacement ring or a kit containing the ring, plunger washer and other washers; sometimes not. In the latter case the entire ball cock assembly must be replaced.

To replace the plunger ring, shut off the water flowing into the tank. Disassemble the mechanism. Remove the plunger and replace the ring. You might as well replace the disk washer at the bottom of the plunger while you are at it.

Toilet tank sweats—This is always due to the temperature difference between the tank filled with cold or cool water and the warm, moist air in the bathroom. The easy way to cure this is to keep the bathroom window open at all times (which is also good in general for the house and its paint). One alternative consists of connecting a mixing valve to the tank's water supply. The other end of the mixing valve (sometimes called tempering valve in this application) is connected to a hot-water pipeline. When the valve is partially opened, you have a toilet tank filled with lukewarm water. There is no temperature differential and the tank stops sweating.

Still another solution is to purchase an insulating plastic liner constructed to fit inside the toilet tank. The tank, being partially insulated, does not become as cold as it did before, and sweating is reduced or eliminated.

Placing a cloth bag around the outside of the tank slows sweating, but in time both the tank and the cloth bag become sopping wet—a solution only if you keep changing bags.

Noisy toilets—All toilets make noise, some more than others, and a few "sing." The unavoidable noise is that produced by the inflow of water. This noise cannot be eliminated, but it can be reduced by replacing a ball cock valve with a diaphragm valve: the newer valve is quieter. Also, sometimes partially closing the shutoff valve leading to the toilet tank reduces the noise, but doing so slows the filling rate, so you have to wait that much longer for the sound to quit and that much longer between flushes.

Avoidable noise is caused by worn ball cock valves. Sometimes they sound like little pile drivers, especially when the water pressure is high (at night, when no other valves are open). The plunger has worn so much and is so loose that it bounces up and down in its cylinder. Try replacing the plunger ring. If that doesn't help, replace the entire unit.

Sometimes the ball cock valve will sing at one particular point in the closing cycle. This is the same trouble, just not as severe. First try reducing the pressure a little by partially closing the feed valve. If that doesn't help, replace the ball cock assembly.

Diaphragm valves are more apt to sing. The trouble is usually due to a sticky push rod. If you follow the rod connected to the float, you will see that it terminates in a pivoted lever, the other end of which presses down on a little rod. This push rod is moved down when the float rises. When the float drops, water pressure on the diaphragm below pushes the little rod upward, at the same time opening the valve. When the rod sticks it vibrates as it ascends and makes all kinds of sounds. Try curing with a drop of oil, using a pair of longnose pliers to work the rod up and down and around (you will have to temporarily remove the lever to do this). When the little rod is loose, the noise will stop (see illustrations 71 and 72).

A constant, low-volume hissing sound is caused by the valve (any type) failing to shut off completely. This is sometimes very difficult to ascertain, as very little water is actually escaping— so little that the water in the tank is up to its mark. What is actually happening is that sufficient water is leaking past the stopper to prevent the ball cock or diaphragm valve from being closed. Neither of these two valve designs shuts off with a snap; instead, they close very slowly. If the water seeps out of the tank just as the ball comes almost up to the closing point, they never close—hence the hiss.

To check for minor water leakage into the bowl, place a piece of dry tissue at the rear of the bowl. If it becomes wet (you shouldn't have the slightest leak at the stopper), clean the stopper bottom and seat. If that doesn't do it, replace the stopper.

Tank takes too long to fill—The cause may be

other water demands within the building. A toilet tank requires water at about 8 psi or more to fill rapidly. If the water pressure at the water service line is low, and if there are too many other faucets open at the time, there simply may not be enough pressure to fill the tank in a hurry (average time to fill should be about 90 seconds). If this is the case, there is no practical solution. On the other hand, there may be any of several other causes not directly related to low water pressure.

The stopper ball may be hanging up. The ball is lifted promptly by the lift wire or chain attached to the control lever. However, it descends by virtue of its weight, which isn't very much. If the lift wire is bent, or if the lift wire guide is corroded, the ball may remain in the open position for a comparatively long time. The tank doesn't start filling until the ball seats itself. In the case of a flap stopper, one arm may be broken, causing the flap to remain open longer than it should: the chain may be catching on something.

When a diaphragm valve has a sticky push rod, filling is delayed. The diaphragm is raised by internal water pressure, which also pushes the push rod upward. If the rod sticks in its guide, the inlet either is never completely opened or is opened very slowly. The flow of water into the tank is reduced and filling time is extended.

The shutoff valve beneath the tank may be partially closed, or another valve in the water feed line may be partially closed. Check by trying the handle.

The feed pipe leading from the shutoff valve to the bottom of the ball cock valve may be partially clogged. This is not uncommon in an old house. Check the pressure at the lavatory faucets. If it is good, check the pressure inside the tank by removing its cover, flushing the tank and watching it refill. If nearby lavatory faucet pressure is good, but inside tank pressure is poor, the 3/8-inch feed line may be clogged. Shut the valve, flush the toilet and disconnect the line. Try blowing through it. If it is gummed up, replace it.

If feed line is clear, shut off main valve, drain the cold-water line and disassemble the shutoff valve. Washers inside these valves sometimes work loose and partially plug up the opening.

Replacing a feed pipe—Should the feed pipe be partially clogged and impossible to fully clear,

or should you damage a feed pipe in the course of working on a ball cock or diaphragm valve (and this isn't at all hard to do), the pipe has to be replaced. Essentially it is a simple operation, but a little tricky in spots.

Start on the morning of a day when you are certain that a plumbing shop or hardware store will be open all day. Close the shutoff valve, then flush the toilet. Back off the large compression nut holding the top end of the feed pipe in place. Back off the lower compression nut. Move the top of the pipe a little sideways to get its top clear of the bottom of the fitting. Then lift the tube up and out of the shutoff valve. (If there is no valve, disconnect the tube from whatever fitting is there.)

Now you need a feed pipe that matches the lower end of the ball cock valve. The newer valves, both ball cock and diaphragm, take a Speedee fitting, described previously. Older ball cock valves take an early form of Speedee fitting. It has a wide mouth, but its washer is flat and fits between the top of the tube end and the bottom of the valve assembly. Usually it is larger than the standard Speedee and is not interchangeable with it.

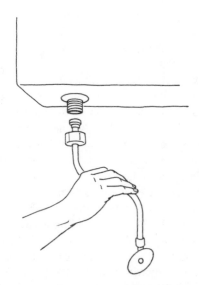

73. When connecting a copper feed tube to a toilet tank or to any fitting or valve, put an S bend in the tube; this will make it easier to connect.

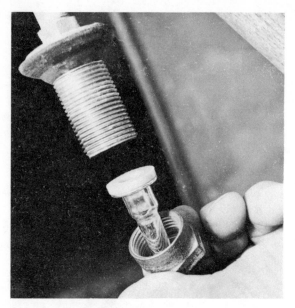

74. Speedee fitting commonly used on modern toilet-tank water supply connections. Large-diameter threaded pipe is the lower half of a ball cock valve.

In either case, the lower ends of the ⅜-inch tube must be inserted into the top of the shutoff valve. (Again, if there is one. All toilets are not the same, and these asides are an attempt to cover as many possibilities as practical within the space available.) This is the nub of the problem: getting the tube end into the opening in the top side of the valve.

More specifically, the Speedee is made in the form of a long, straight tube (see illustrations 149 and 166). The bottom of the toilet valve and the top of the shutoff valve are always offset. Therefore the Speedee tube must always be cut and bent to fit. In doing so, bear in mind that the tube must enter the shutoff valve opening vertically for about ¾ inches and that the top of the tube must be in line with the bottom end of the valve assembly. If these conditions are not met, the joints will probably leak when tightened.

Start by bending the tube carefully to shape. You can do this with your hands if you take care not to make the entire bend at one point on the tube. Bend it a little at a time, moving the position of your hands as you do so. Connect the

upper end of the tube after you have bent it to shape. Rotate the tube until it is alongside the shutoff valve. Mark the point at which the tube should be cut. (Remember, it enters the fitting about ¾ inch.) Remove the tube, cut it and carefully remove the inner and outer burrs.

Slip the upper compression nut onto the tube. Follow with the lower compression nut and metal ring. (Use a plastic ring if you can get one.) Insert the lower end of the tube in the shutoff valve. Push the ring down and just start the compression nut. Rotate the tube until it is in line with the upper valve end. Slip the washer in place. Raise the upper compression nut. Tighten it and then tighten the lower nut and the job is done. When the tube is correctly bent and then cut to the correct length it will almost slip into place by itself. If you have erred, correct the mistake by bending the tube along its middle.

When the bottom of the toilet tank is very close to the shutoff valve, it is difficult to install a new water feed tube because there isn't sufficient tube length in which to make satisfactory bends by hand. One solution is to make a complete loop in the feed tube.

Toilet won't flush properly—This can be caused by a partially plugged drain (drains and their problems are covered in Chapter Eleven), insufficient water or a low rate of flow.

Remove the tank cover and note the height of the tank's water level when the tank is full. If it isn't up to the mark, or about 2 inches from the top of the tank, adjust the float until it is. Even the lack of a few inches of water height seriously decreases the toilet's ability to flush.

If the tank's water height is correct, the trouble may be due to an obstruction in or near the pipe leading to the bowl. Operate the toilet a few times and see whether or not the stopper is raised clear of its seat. If it isn't, the fault may be due to the position of the chain or lift wire on the control lever arm. Move the chain or wire to a hole farther away from the lever pivot. Doing so increases the upward motion of the chain or wire when the control lever is operated. There is also a possibility that the guide may be set too low, preventing the stopper from being raised high enough. And there is also a chance that the stopper ball has a leak that prevents it from floating properly. In the latter case the stopper is raised high enough, but sinks too quickly into its seat, preventing all the water in the tank from leaving and so reducing the toilet's ability to flush soil out of the bowl.

CRACKED TOILETS

There isn't much you can do about a cracked toilet bowl because it is very difficult to get it dry enough to make epoxy cement stick. But toilet tanks and toilet tank covers are easily repaired with epoxy cement.

In the case of a broken cover, remove the pieces and let them dry thoroughly. Apply the cement per instructions to the broken edges and reassemble. When the cement is hard, back up the crack(s) with a sheet of aluminum or hard plastic. Then apply a generous quantity of the epoxy cement to the porcelain and the backup material. When the cement hardens, return the tank cover to service.

Tank repairs require a little more effort. The water to the tank has to be shut off and the tank must be dried out completely. This can be done with a sponge or a rag. The tank should be given several days in which to dry thoroughly. If this isn't practical, use a propane torch and carefully dry it with that. When the tank is dry, apply the epoxy and reinforce the joint by cementing a thin plate of plastic or metal across the crack.

In the case of both the cover and the tank, do not let too much time pass after the occurrence of the crack before you patch it. If you do you will find that the crack has filled with one or another mineral (rust) and that the epoxy may not stick. Toilet removal and replacement is discussed in Chapters Eleven and Twelve.

TOILET SEAT REPLACEMENT

The only difficulty you will encounter in replacing an old toilet seat is that of removing it. With time and moisture the nuts rust onto the bolts holding the seat hinge. Generally there isn't too much room to get a wrench on these nuts. If that is the problem, do not force the wrench into the tight space: you might crack the porcelain. The better approach is to hacksaw the bolts off. Slide the edge of the saw blade under the top of the bolt above the porcelain. The blade will slide on the porcelain without harming it very much.

New toilet seat assemblies usually have bolts that can be moved toward and away from one another so that most new seats fit all toilet bowls, but to make certain you might measure the hole spacing.

Loose screws—Toilet seats are now most often made from a kind of pressed paper composition. When the screws are pulled out it is almost impossible to get them to take hold again. Try filling the old screw holes with epoxy, drilling new starter holes and then running the screws in again. It sometimes works, but not always.

CHAPTER EIGHT

Fixing Leaking Faucets and Valves

While it may appear as if there were dozens of totally different valves and faucets in use today, it is not so. Actually there are only four basic types: **gate, compression, slide** and **Fuller.** Differences in appearance result from minor operational variations and external design choices in handles, spouts and the like. The following suggestions for the repair of the various basic types are applicable to all variations of that type. Whatever you have to do differently from the procedures described in the text will usually be obvious once you go to work on the faucet or valve.

BASIC TYPES

As stated previously, a faucet is a valve designed for use at the end of a run of pipe. Valves are designed and used somewhere along the length of the pipe run. Again, as stated, the four basic types are gate, compression, slide and Fuller.

Gate valves—You can recognize a gate valve by its operation and appearance. To fully open a gate valve from a fully closed position it is necessary to turn its handle half a dozen times or more. While turning you will feel no change in handle response. If you didn't know it was a gate valve you would imagine that nothing was happening, that the valve was broken inside. If you look at a gate valve alongside a compression valve you will see that the **bonnet,** the lump of metal below the handle on the gate valve, is much larger (higher) than that on the compres-

sion valve. The larger bonnet is necessary to permit the gate to be moved completely free of the internal valve opening.

A gate valve is literally a gate. When you turn the valve's handle you are lifting or lowering a tapered wedge of metal (gate) from between two holes leading to the two pipe ends joined to the valve. When the gate is all the way up, there is practically no obstruction to the flow of water. When the gate is closed, all flow is shut off. Partially opening a gate permits partial flow, but the

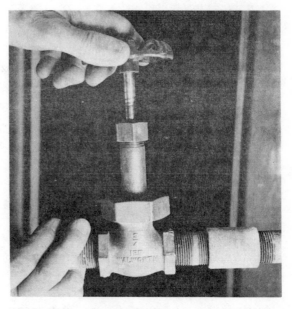

75. One form of gate valve.

movement of water over the carefully machined gate erodes its surface. Therefore a gate is never operated in a partially opened (or closed) position. Once the surface of the gate has been worn down, the gate will never make a perfect seal again—it will always leak. A repair is a machine shop operation. The entire valve must be removed and the gate and its seats ground down and made perfectly flat again.

Gate valves are several times more expensive than compression valves of equal size and quality. When they are properly used (fully opened or closed), they last practically forever.

In the home, gate valves are used as the main valve positioned in the water service line just ahead of the water meter. The expensive gate valve is used there because, when fully opened, it restricts the flow of water less than ⅕ as much as an equal-size compression valve. Since all water used in the home must flow through the main valve, even a little flow restriction can result in considerable loss of pressure. (Maximum pressure loss occurs with maximum flow.)

Gate valves are also used in home heating systems in both hot-water and steam lines for the same reason.

Compression valves and faucets—You know that you are operating a compression-type valve or faucet when you can open and close it with two or three turns of its handle.

Compression valves are so named because turning the handle to close the valve acts to compress (press) a washer against a hole leading to the pipe feeding water into the valve or faucet.

Compression valves and faucets wear a little each time they are used, because the washer rubs against the seat as the handle is turned. (Some of the newer designs feature greatly reduced washer wear.) Therefore, it is normal for a compression valve and faucet to require a new washer every few years or so, depending on the amount of service it sees.

Since faucets are designed for almost daily use, they are constructed to sustain this wear and to be more or less easily repaired. Valves not intended for frequent use frequently cannot be repaired very much beyond washer and packing (to be shortly explained) repairs.

Most of the valves and faucets used in the

76. A small compression valve made to be joined to copper tubing by sweating.

home are of the compression type. You will find them used for shutoff in feed lines connected to sinks, lavatories, tubs, showers and toilets. The ball cock valve discussed in the chapter on toilets is a type of compression valve.

Slide valves—This type of valve is used in single-lever water controls. Most new homes come equipped with a single-lever faucet on the kitchen sink, and some have them in their lavatories.

The single-lever faucet is easily recognized (see illustration 107). You open, close and mix hot and cold water with a single lever. The swing of the lever, right and left, selects water temperature. The angle of the lever controls flow rate. On kitchen sinks the lever is comparatively long. On lavatories the lever is no more than a few inches in length.

The single-lever faucet has a number of advantages over the compression faucet. Although pressure is applied to both the compression washer and the slide washer or pad, the compression pressure is very high, whereas the slide pad's pressure is comparatively low. Thus slide pad wear is far less than that of the washer. The single lever also affords more rapid control. Thus with a compression faucet you waste a few seconds opening and closing the valve. The same can be accomplished in a fraction of a second with the lever. It doesn't appear to be important, but over the year you can save a considerable

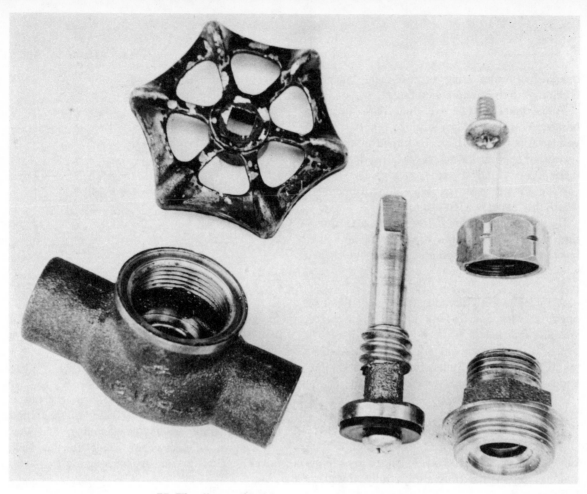

77. The disassembled parts of a small compression valve.

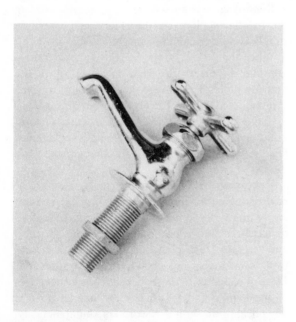

78. Typical compression faucet. Flange rests on surface of sink or lavatory. Nut holds faucet in place. Water connection is made through hollow shank (threaded portion).

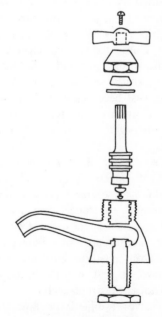

79. Interior of typical compression faucet.

quantity of water using a single-lever faucet as compared to the compression faucet.

Fuller faucets—You will find these in older homes. You can easily recognize this type when you turn its handle. It can be turned in either direction to open and close. A half turn does it and there is no "solid" closed position. The handle can be turned past the fully closed point, to where the faucet is open again.

In a sense the Fuller faucet is also a compression faucet. Turning the handle pulls an inner lever which pulls and compresses a cone-shaped washer against an opening leading to the feed pipe.

A Fuller faucet's advantages are that it can be operated very quickly and that it is in place, meaning that unless you are going to completely overhaul the kitchen or bathroom, there is little point in replacing a Fuller faucet with a similar-appearing, old-fashioned compression faucet.

EXTENDING FAUCET LIFE

As previously mentioned, the life of a gate valve can be immeasurably extended by simply making certain that it is always fully closed or fully opened.

The life of a compression valve or faucet can be extended by never closing the valve unnecessarily tight. (If a lot of pressure is required to stop a drip, the washer probably needs to be replaced.) Continued forced closure quickly destroys the washer and damages the seat. As the washer makes its seal by resting against the seat, a perfect seal becomes impossible when the seat is scored. At this time the faucet will leak no matter how tightly the washer is pressed against the seat.

Another "never" is never leave the faucet or valve slightly open. A drip will wear a groove across the seat sooner than you might imagine. Once grooved, the faucet or valve can never be completely closed again without repairs.

Fuller faucets are also damaged by partial closure. Drips past the ball and across the seat will groove the seat, and the seats of Fuller faucets are not always replaceable.

Slide valves (single-lever faucets) are rarely left in the drip position because it is so easy to

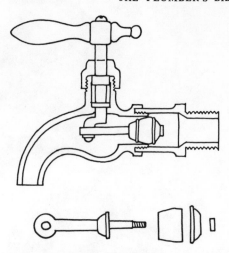

80. Basic parts of a Fuller faucet. Excessive wear at stem and control-rod joint reduces pressure on valve and eventually makes faucet useless.

move the lever all the way and close the faucet completely. However, this ease of operation sometimes leads to handle-slamming, and that can damage the mechanism. Except for this, slide valves need no special care for long life.

REMOVING HANDLES

Removing screws—When the screw holding the handle in place on the stem can be seen, it simply can be backed off and removed.

When the screw cannot be seen you will generally find it beneath the snap cover on top of the faucet handle. The cover often carries an *H* or a *C* (hot, cold). The better-made handles appear to have covers machined in place. Try a sharp knife and pry very carefully. Single-lever handles are removed somewhat differently and are discussed later.

Removing the handle—This can be the hardest part of the entire job. Sometimes the handles are easily lifted off. Sometimes when the faucet is old or has been wet a long time the handle corrodes and "freezes" fast to the stem. In such cases, use two screwdrivers as shown in illustration 84. Place the points of both beneath opposite edges of the handle and try to pry the handle up and off, pushing on both screwdrivers simultaneously. If this doesn't do it, there is a special handle-

81. To get at the screw holding the handle down on this old-timer, the top is unscrewed.

83. Once the handle screw is uncovered it is backed out and removed.

82. To get at the screw on this modern faucet, the cover cap is pried off with a sharp instrument.

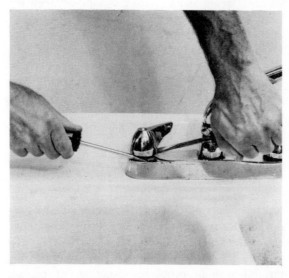

84. When a handle is stuck, try using two screwdrivers as levers to pry it off. If this fails, you will have to purchase a special handle-removing tool. Plumbing supply houses carry it.

removing tool sold in plumbing shops. It is advisable to purchase this tool rather than to attempt to force the handle, since it is easy to damage the porcelain fixture while banging on the handle with screwdrivers or other makeshift pries.

STEM LEAK REPAIRS

All valves and faucets except single-lever faucets have stems, from around which water will leak in time. As the designs of all stems are identical, their repairs are identical and are grouped here. (The exception is the single-lever faucet, which will be discussed later.)

Description—The top end of a valve or faucet stem is connected to a handle. The lower end carries the washer or gate. The stem itself is coarsely threaded so that when it is turned it moves in and out of the body of the valve or faucet. When the valve or faucet is opened, the stem is raised. Water enters and the lower end of the stem is submerged in water under pressure. To permit the stem to turn and move up and down the stem is "packed" in a soft, lubricated material (except in the Fuller). Called **packing**, the material is held in place by a compression nut called a **packing nut.** The packing nut is obvious on a valve and some of the older faucets. On newer faucet designs the packing nut is hidden by the handle. When the handle is removed it is the first nut on the stem.

Minor stem leaks—With time and use the packing wears and a space develops between the stem and the packing. When the valve or faucet is opened, water now emerges from the space between the packing nut and the stem.

To cure this, the packing nut is given a half turn or so. Doing so compresses the packing. The gap is closed and the leak is stopped.

With the passage of more time and more packing-nut turning, there will come a point when the packing nut strikes the body of the valve or faucet and can be tightened no more. At this time the old packing must be removed and replaced.

Removing the packing nut—When the valve has been correctly installed (faucets usually cannot be incorrectly installed), closing the valve closes the water inlet side of the line and you can remove the valve handle and back the packing nut off without problem, assuming, of course, that there is no water on the "out" side of the valve (which would be the case with a valve installed in a basement). When a valve has been incorrectly installed, water under pressure will emerge when you back the packing nut off, whether the valve is opened or closed. In this case, the packing nut is retightened and the system is closed and drained before proceeding.

Replacing the packing—On older faucets and valves the packing consists of a cotton thread,

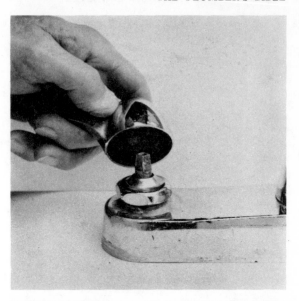

85. The packing nut on this faucet is found underneath the handle.

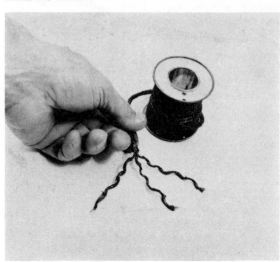

86. Thread packing in commercial quantities is fairly thick, but easily unraveled for use around close-fitted faucet and valve stems.

waxed and covered with powdered graphite, wound around the stem. On newer valves and faucets the packing is in the form of a thick, soft black or gray washer, similarly lubricated. In either case the packing nut is backed off and removed or pushed up and out of the way on the stem. A small, thin screwdriver or a piece of stiff

87. Two examples of preformed, washer-type packing.

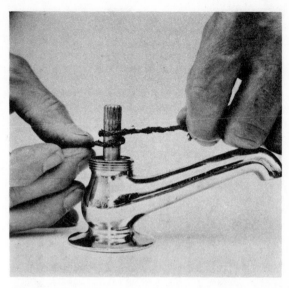

89. How thread or string packing is wound around a faucet (or valve) stem.

88. Washer-type packing used with a small compression valve.

wire is used to remove all the packing, including the bits and pieces from the space between the stem and the inside of the valve or faucet body. You want this space clean.

If you can secure a replacement packing washer, fine. Remove the handle, if you have not already done so. Slip the washer packing over the stem and down into its niche. The packing nut is made snug enough to hold water, and that is it.

If you cannot secure a packing washer, or if the faucet uses string packing, replace it with new packing string (plumbers simply call packing string "packing"). There is nothing critical in the selection of the string except to make certain that it is thin enough to permit several turns around the stem. One fat turn of string, its ends butted up against each other, will not do. If the packing you purchase is too thick, simply unravel it.

Should packing be unavailable at the moment of repair, you can use cotton string heavily waxed as a substitute.

The quantity of string you wrap around the stem isn't critical either, but do not use so much that you cannot get the packing nut easily started. As with the packing washer, the packing nut following the thread is made just tight enough to enable the packing to hold water.

New packing doesn't cure—When a valve or faucet stem is bent or very worn, new packing alone may not stop a stem leak, or may not last very long. If you can clearly see that the stem is bent as it enters the valve or faucet, don't waste time with new packing. Get a new stem.

If the stem is so worn you can shake it from side to side as soon as you open the faucet a crack, the same advice pertains. New packing will not help stop this stem from leaking.

90. To remove the valve-assembly portion of a modern faucet, back off and remove the large nut exposed by removing the handle.

91. An O-ring valve assembly. The two rubber rings seal the stem. The spindle (between the fingers) fits inside the tube to the left, which screws into the faucet body. The entire assembly is sold as a replacement part.

New stems—Stem kits for various popular faucets are available in plumbing supply shops. Sometimes a new stem and new packing cures the trouble. Many times it doesn't. The balance of the mechanism is worn too much. Most often the replacement stem does the trick for a few months and then starts to leak. Faucet replacement is discussed in Chapter Twelve.

WASHER LEAKS

Leaks from around a faucet or valve stem are separate and distinct from leaks past the washer. The only time they are related is when there is a bad washer leak and a stem leak, in which case the faucet or valve will leak in both places simultaneously.

Cause—Faucet washer leaks are visible as drips emerging from the faucet spout. Sometimes you can hear a hissing sound. Valve washer leaks are not readily discernible, as a valve is in the middle of a pipeline. You learn of valve leaks when you go to shut off a pipeline and find that it cannot be done. This usually happens when you go to repair a faucet and find that water continues to come out of the pipe despite the closing of the supply valve.

In any event, compression-faucet leaks and compression-valve leaks are cured in the same way, so we will stop mentioning valves from this point on.

As stated, a compression faucet leaks when the washer does not make *perfect* (and bear that word in mind) contact with its seat. Imperfect contact can be caused by a worn washer and/or a worn, rough or grooved seat. Washers wear with normal use. When a worn washer is not replaced in time, three things happen: more and more pressure must be used to close the faucet, thus accelerating the destruction of the remaining washer surface; the washer and/or its support chews into the seat, abrading its surface; water flow is never completely shut off, resulting in a trickle, which in far less time than you can imagine will cut a groove across the washer seat.

Seats normally last for many years if the washers are replaced when necessary and if the faucet is properly closed after each using. Let the faucet drip and the seat becomes grooved.

93. How a washer fits into the stem of a valve spindle.

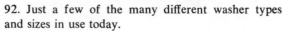

92. Just a few of the many different washer types and sizes in use today.

Removing the washer—The pipeline is shut down and drained. The faucet handle is turned to open the faucet and raise the handle upward from the faucet body. If the faucet stem is short, the handle is removed. If the stem is long, the handle can be left in place. Next, the packing nut is backed off. The stem is given an additional turn or two and removed from the body of the faucet.

The washer, or what remains of it, rests in a shallow, circular cup attached to the lower end of the stem. The single small bolt holding the washer in place is removed and the washer is pried out. Any bits of rubber adhering to the metal are scraped off.

Selecting the washer—Although this sounds easy enough, it isn't due to the unremitting efforts of all the faucet manufacturers to make entirely different faucets and washers almost every year and to make their faucets and washers different from everybody else's. As a result there are two basic types of washers and at least a dozen washer sizes in *popular* use today.

If there is sufficient washer remaining to give you a clue, decide whether it was flat or hemispherical and select a similarly shaped washer from the box of assorted washers and screws sold in most hardware shops and plumbing supply shops. Pick a diameter that is slightly smaller than the flat cup it fits. Both the washer's shape (flat or hemispherical) and diameter are critical. Install the wrong one and your faucet will drip.

The diameter of the washer must be slightly less than the provided space: about 2 hair breadths. If it is not it will wrinkle when you force it into place and when you tighten the small bolt that holds it. A flat washer usually will not make a tight fit against a seat shaped to accept the ball-bottom washer, and vice versa. The choice of washer types also determines, to a small but sometimes important extent, the height of the stem above the faucet body. Use a flat washer where the round is needed and you may find the handle striking the sink before the faucet is fully closed. Still another tip: if you use an undersized ball or hemispherical washer it may enter the valve seat opening and remain fast when the handle is turned. This will rapidly loosen the bolt holding the washer and both will fall off the faucet's stem.

The flat washer is positioned with the printing against the stem's end. The curved washer rests with its flat side against the stem's end.

If there isn't sufficient old washer left to deduce its original shape, try a flat washer if the faucet is old and a hemispherical washer if the faucet is relatively new.

There are several different kinds of washers for sale today. The cheap washers made out of pressed paper and colored gray for cold water and red for hot aren't worth using. Purchase the black rubber kind, or the sandwich type made of three layers of plastic. There is also a flat washer mounted in a little brass disk that rides on a pair of legs. The legs slip into the bolt hole. The washer rotates in its disk. This design purports to eliminate washer wear due to the washer rubbing against the seat as it is closed.

Having selected your replacement washer, examine the bolt that held the old washer. If it looks somewhat rough and red and granular it is ready to disintegrate: replace it with a new bolt. You'll find a selection in the box holding an assortment of washers. Choose a bolt that just touches the bottom of the hole as it presses lightly down on the washer. This is the optimum bolt length. If that isn't to be found, find a shorter bolt and just make it snug. If you make it too tight you will crimp the washer and the faucet will drip.

Examine the seat—Use a flashlight and peer down the hole whence the stem came. You will see a circular boss, a shiny ring of brass with a square or octagonal hole leading outward to the faucet's spout. If you do, all is well and you can reassemble your faucet. The job is done. If not—if the ring is corroded, or if it is bright but has a dark line or two crossing it—the seat is defective and must be either replaced or refaced. Refacing is a minor machining job that you can do with a special hand tool.

Removing and replacing the seat—Poke either end of your seat-removal tool down into the center of the seat. One or the other end will fit. If neither does, there is a chance that the tapered end of the tool may be too narrow. If so, hacksaw a little off the appropriate end and try again. When the tool is in place, give it a light tap with your wrench to make certain that it fits tightly. Then use your crescent wrench and turn the tool counterclockwise, using a steady pressure. The seat is a small, shallow, threaded ring of brass about ¾-inch across and about ⅜-inch high. If it falls off the tool as you try to lift it out, use a hooked wire to retrieve it. If that doesn't work, open the water line and let the water wash it up and out.

There are more than 24 slightly different faucet and valve seats used on American faucets today. They cannot be interchanged because they have different threads, though a little difference in height and shape can be accommodated. The best way to find the one you need is to bring the old one to a plumbing shop that carries replacement seats and has a test board (with 24 test holes). Try the old seat in all the holes until you find the

94. The seat at left is useless. It has been worn through (dark line) by a constant water drip. Seat to the right is a new, replacement seat.

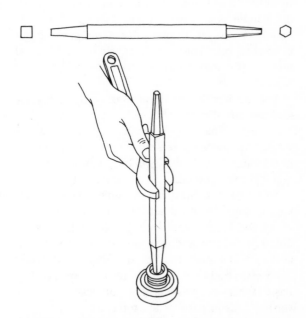

95. How a straight-bar type seat-removal tool may be used to remove a valve seat. Use either the tool's octagon or square end, whichever end fits the seat best.

96. Two dissimilar seats. The one at left is old. Note that you have to match seat diameters, seat heights and seat threads to fit a new seat into a faucet or a valve.

one that works best. Note that some seats can be *forced* to go into several test holes. Markings on the test board will identify the correct replacement. If you cannot get the correct replacement at the first shop, continue looking. If you force a seat into place, very likely the seat will leak and the threads will be damaged. In which case, you will be forced to replace the faucet; the correct seat will no longer fit properly.

When you have the correct replacement, apply some pipe dope to the threads and to the tool (to make the seat stick). Slip the seat on the tool end, invert the tool and carefully lower the seat and start it. Do this very carefully, backing up once or twice to make certain you haven't crossthreaded the seat (started it at an angle). When you are certain that it is properly started, screw it home and make it snug. It doesn't have to be very tight to hold water.

Reassemble your faucet and you're done.

Refacing the seat—When someone has turned the square opening in the seat into a circle by trying to remove it with a screwdriver, or when the seat is integral with the faucet body and obviously cannot be removed, the seat has to be refaced if it is rough or grooved.

A satisfactory refacing job can usually be accomplished with a professional refacing tool. It cannot be done with most of the five-and-dime tools sold for the purpose. The word "most" is used because some of the cheap refacing tools do work well.

The professional refacer sells for more than ten dollars today. It can be used with a wide range of faucet and valve types and sizes. The cheap refacer sells for under three dollars. The type that uses the faucet packing nut as a guide usually works satisfactorily. The type that must be guided by hand is a waste of time because it is impossible to keep the tool perpendicular while you are turning it.

All three types look somewhat alike. The best is like a long-stemmed letter *T*. The other two are like short-stemmed letter *T*s. All have their cutters fastened to the bottom of the *T*. The best has a selection of cutters and guides which can be adjusted to fit the individual faucet. The unusable type has no guide. The good but inexpensive type is the best deal financially when it fits your faucet. Unfortunately, it doesn't fit many faucets.

All of the refacing tools are used the same way. The faucet is disassembled. A cutter matching the seat is selected and affixed to the end of the tool. The tool and cutter is inserted into the faucet until the cutter just touches the seat. The guide, if there is one, is locked to the faucet. The tool is rotated. The cutter removes a slice of the seat. The tool is removed and the seat is examined. If the one slice didn't do it, the operation is repeated again and again until the seat is one smooth, bright, unmarked circle.

Refacing the seat reduces seat height. When so much seat is removed that the faucet handle strikes the faucet body or packing nut, you are in trouble. Sometimes the bottom of the handle can be trimmed a little with a file. Sometimes you can solve the problem by trimming the top of the packing nut. Sometimes installing a hemispherical washer in place of a flat washer will give you the fraction of an inch you need. Sometimes you can get away with using two flat washers. When none of these gimmicks works, there is nothing for it but to replace the faucet or use a special seat.

Seats within seats—When you do not want to purchase a refacing tool, or when the seat has been refaced too many times and is now too low, you can sometimes screw a special nylon "cover" seat into the old seat and solve the leak problem. There are several drawbacks to this solution, as with many other "simple" solutions. Screwing the new seat into the old one reduces

97. A professional valve-seat refacing tool in use.

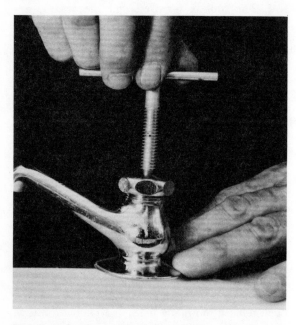

98. A low-priced valve-seat refacing tool in use. They are comparatively inexpensive and work fairly well when they fit the faucet. Note that they usually have a cutter suited only for a flat washer.

the rate of water flow, which isn't an impossible hardship. Worse is that the added seat raises the faucet stem so that internal clearance is greatly reduced. As a result, many faucets cannot be opened or opened fully with the second seat in place. Still another problem is that the cover seat is not carried by very many shops.

REVERSE COMPRESSION FAUCETS

What may be termed the **standard compression valve and faucet,** because it is the type in greatest use, operates by compressing the washer downward against the seat. The stem moves into the valve body or faucet when you close it. The **reverse compression faucet** operates in the reverse direction. The stem is raised when the faucet is closed. This is the only difference that can be seen from the outside.

When you take a reverse compression faucet apart, however, you will find that it is different in a number of ways. The bottom end of the spindle is cone shaped, much like an automotive engine valve. Closure is effected by bringing the metal cone up and against a flexible-composition seat ring. The stem is sealed by a packing ring, which is usually a simple O-ring that fits inside a sleeve.

Stem leaks—Stem leaks in a reverse compression faucet are caused by a worn O-ring. Replace it and your problem is solved.

99. A disassembled reverse compression faucet valve. Water flow is shut off when stem is raised against seat ring. *Courtesy of Crane Company.*

Drips—Drips are caused by a worn seat ring, which acts as a washer. Replace the seat ring and your leak is stopped, unless you hesitated too long, in which case the seating surface on the cone end of the stem has become grooved (like the metal seat in a standard compression faucet). When this happens the stem has to be replaced.

SHOWER CONTROLS

Showers usually have three control knobs. The two outside handles control the flow of hot and cold water. The central knob, if there is one, determines whether the water emerges from the tub's spout or the shower head.

The outside controls are compression valves, very similar to the standard compression valves and faucets shortly discussed. The only difference is that they have long stems which are necessary to pass through the wall thickness. The shower control casting is fastened to a cross brace behind the bathroom wall. The central control is a simple directional valve which almost never breaks down, so it requires no description here.

Shower valve problems—Shower valves leak either at their stems or past their washers. Stem leaks are usually not immediately visible, as the emerging water drips down between the walls. Sometimes the drip can go unnoticed for years. The wood and plastic absorb the water and re-main moist, rotting slowly. Sometimes the wallpaper downstairs becomes loose, and sometimes the ceiling paint becomes discolored. Any sign of moisture in the floor below the bathroom shower is reason for examining the shower controls. Leaks past the washers, on the other hand, usually drip out the tub spout.

Disassembling shower valves—Remove the handle. Remove the escutcheon or flange. In some bathrooms the stem is hidden within a chrome-plated tube. This tube must be unscrewed on some valves and simply pulled out on others. Use tape wrapped around the tube to prevent marring it should you have to use a wrench. Removing the handle, tube and/or escutcheon will reveal the valve stem, the packing nut and the top of the valve body, which will be in the form of a nut.

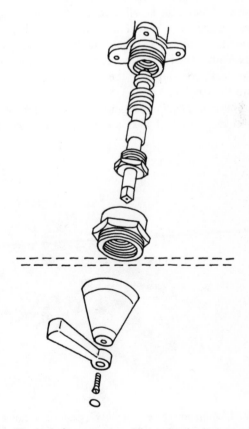

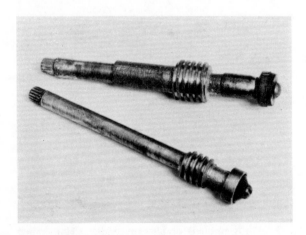

100. Two shower control stems. Arrangement of parts is the same as a conventional sink compression faucet.

101. Typical shower valve. The valve body is a casting which also carries the second valve and is threaded to accept pipe leading to the shower head.

To get the packing nut off you need an automotive-type, deep-socket wrench of the correct size, plus a small Stillson wrench. The socket wrench is slipped over the stem and the Stillson wrench is used to turn the socket.

To get at the washer you need a larger size automotive-type, deep-socket wrench. This slips over the nut-shaped top of the valve body. Backing this nut (and body half) off permits you to remove the top half of the valve.

From this point on, repair procedure is similar to that used with ordinary compression valves. Just be careful when reassembling that you don't cross-thread the valve body, and don't cross-thread the plated tube.

DIAPHRAGM FAUCETS

A diaphragm faucet is a type of compression faucet. The seal is effected by compressing a washer against a seat. However, the washer is a hat-shaped rubber part that doubles as the packing. Thus, when you have either a stem leak or a seat leak you replace the diaphragm (hat) to cure the trouble.

You can usually recognize a diaphragm faucet from the outside by its short, fat body and the fact that it can be closed or fully opened with less than one full turn.

Disassembling a diaphragm faucet—Pry the cap off. Unscrew the machine bolt and remove the handle. Use a crescent wrench to back off the large nut that is revealed. When this nut is removed you will find a short tube that holds the threaded stem and the stem itself. The lower or inside end of the stem may or may not carry an O-ring, plus a thin metal washer and the hat-shaped rubber diaphragm. Look into the faucet body and you will see the conventional seat.

Stem leak repairs—Replace the O-ring if there is one. If that doesn't do it, replace the diaphragm as well.

Washer leaks—Leaks past the washer are caused by a worn and/or tattered (leaky) diaphragm or a defective seat. You can check the seat by simply examining it. If it is rough or grooved, replace it. If the diaphragm is defective replace it, making certain to place the thin, flat washer beneath the diaphragm (between it and the stem). The metal washer aids the diaphragm in turning, which is one reason why this type of washer lasts so long. It turns when it contacts the seat and is not abraded when the faucet is closed, as is the washer on a conventional compression faucet.

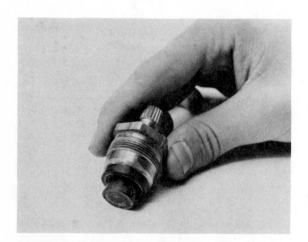

102. Diaphragm-valve assembly (part of a faucet). Like the O-ring assembly, the entire unit screws into the body of the faucet and can be replaced in part or as a unit.

103. Disassembled diaphragm valve. Hat-shaped washer is at left, atop the short stem. This design (American Standard) has a narrow O-ring just under the washer.

Faucet opens by itself—The threaded spindles of these faucets are often heavily lubricated. Sometimes when making a repair a little of the grease gets on the diaphragm. When this happens there is no friction between the rubber diaphragm and the valve seat. The natural expansion of the rubber against the valve stem causes it to open a crack. Try disassembling the faucet and wiping the seat side of the diaphragm clean.

DUAL FAUCETS

Dual faucets consist of two faucets or valves that feed into a common spout. These valves are serviced individually and exactly as they would be if they were not mounted on a common casting.

Stem leaks—If opening the faucet on one side causes the faucet on the other side to leak around its stem, the trouble is a leak at the valve stem of the latter faucet. If either stem leaks when the faucet incorporating it is in its open position, the problem is in the leaking stem.

Swivel leaks—Leaks at the junction between the swivel spout and the dual faucet body are caused by worn O-rings. To cure this, wrap tape around the nut that holds the swivel spout in place. Remove the nut and lift the spout off. This

105. O-ring used to seal spout to dual faucet body.

will reveal the one or two O-rings used there. If the rings are worn or broken the joint will leak. Replace the rings, making certain to get the correct size ring. The correct size ring will fit with just a little tension. If it is loose or if it has to be stretched for installation it is the wrong size.

SINGLE-LEVER FAUCETS

The flow of hot and cold water from a single-lever faucet is controlled by slide valves, which were briefly mentioned a few pages back. Two designs currently provide the basis for most, if not all, single-lever faucet and shower control manufacturing. One uses a flat-slide valve. The other uses a ball. Their actions, however, are identical. Service is somewhat different, but not tremendously so.

BALL-SLIDE DESIGN PROBLEMS

Ball-slide design—A suitably pierced ball, attached to a short rod, rests on spring-loaded pads which lead to the hot- and cold-water supply. The position of the holes in the ball in relation to the pads and a third hole leading to the spout determines hot- and cold-water flow, volume and

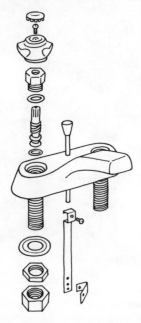

104. Exploded view of one dual faucet design.

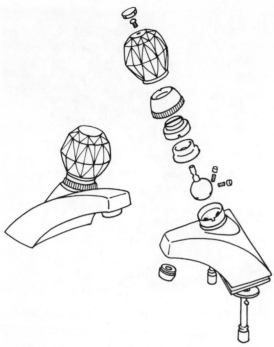

106. One type of ball-slide single-lever faucet. Adjusting the ring varies the pressure on the packing, controls stem leakage.

shutoff. The ball design differs from the flat mainly in that the ball has a cone-shaped packing ring that seals it to its support, and the pads which act as washers are held in place by springs. The pads are not always easy to replace.

Stem leaks—Water that appears around the handle undersides and not at the spout is a stem leak, although this type of faucet has no stem. In the ball-design faucet stem leaks are usually due to a loose adjusting ring **(cap)** or worn packing. On some models you can get at this ring and tighten it by hand without removing the control handle. On others, the handle must be first removed.

To remove the handle it is tilted upward and an Allen wrench is used to loosen the set screw you will find there. Now the handle can be lifted up and free of the post it rides on.

Give the ring just below the handle as many turns as necessary until you either stop the stem leak or make the control lever (when you replace it) too tight to move. If reasonable tightening doesn't stop the flow of water from beneath the handle, the packing is probably defective and must be replaced. This is how it may be accomplished.

107. To remove the handle on most single-lever faucets, use an Allen wrench to loosen the set screw you will find underneath.

108. Remove the adjusting ring to get at the valves in a ball-slide single-lever faucet.

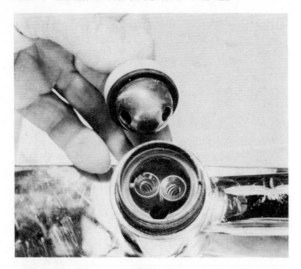

109. Lifting the ball exposes the valves (pads) and the supporting springs.

Close the hot- and cold-water supply lines. Remove the handle and unscrew the ring. The packing, which is a molded ring, is then lifted off and replaced. Generally only an identical replacement part will work here. When you reassemble the faucet take care to get the cam back into its proper place and make certain that you do not tighten the ring pressing down on the packing any more than is necessary.

Spout leaks—This may be caused by a loose adjusting ring and/or worn pads. Shut off the water lines, dismantle the assembly and inspect. If the pad or pads are worn, the spring and its associate parts are probably thin and weak. It is advisable to replace them all at the same time. Most often, you are forced to purchase an entire parts replacement kit and cannot purchase an individual pad.

If the ball itself has been scored by a spring that got between it and the housing, new pads usually will not stop the drip and you have to replace the entire assembly or even the entire faucet.

FLAT-SLIDE DESIGN PROBLEMS

Flat-slide design—The two feed pipes and the spout are connected to a thick, three-hole, flat brass disk. A stubby porcelain cylinder is bolted atop the disk. The cylinder has three holes lined with special rubber grommets. A second piece of porcelain rides atop the first and is coupled to the control lever. The second piece of porcelain has slots on its underside. When the lever is moved so that the slots are evenly positioned over all three holes, hot and cold water flow up the two small-diameter holes and down the large hole and out the spout. When the slots are aligned with only one small hole and the large spout hole, cold or hot water alone flows out of the spout. When the top piece of porcelain is moved to one side, the small holes are sealed and no water flows.

Stem leaks—Water that appears around the handle underside and not at the spout is termed a stem leak, even though this type of faucet has no stem. Such leaks can be caused by a loose slide, a worn slide, a cracked slide or worn rubber grommets. To check these the handle and cap must be removed.

110. Partially disassembled flat-slide design single-lever faucet (American Standard). The porcelain disk between the plumber's fingers carries the rubber grommets that act as a valve. Pressure of disk against grommets and in turn against lower brass disk keeps water from leaking out.

Lift the control handle. Use an Allen wrench to loosen the set screw you will find there. Now lift the handle assembly up and free of the square post it rides on. Unscrew the plated brass collar.

Examine the joint between the porcelain and the brass. If water is seeping from this joint, try slightly tightening the two brass bolts to either side of the short square rod. You cannot tighten these bolts all the way, or it will be difficult or impossible to operate the control lever.

Examine the porcelain for a crack. If there is a crack, shut off both the hot- and cold-water lines. Remove the valve assembly. Remove the rubber grommets. Dry the porcelain in an oven at low heat and then try epoxy cement. If that doesn't work, the entire assembly has to be replaced.

If there is no crack, and if the bolts are fairly tight and there still is a leak from between the disks, shut off both water supplies and remove the two bolts. Turn the porcelain upside down and examine the rubber grommets. If they are not smooth and unworn they must be replaced. Examine the top of the brass disk. This surface must be perfectly smooth and unworn. If it has been corroded or grooved, its surface must be machined

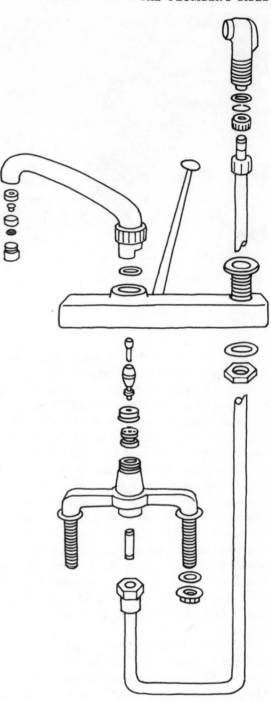

111. Underside of porcelain disk. Note the three grommets. When they are worn the faucet will leak. When the brass plate on which they ride wears, it must be resurfaced or the entire faucet must be discarded.

112. You will find the diverter valve inside the body of the faucet, after you remove the spout. The valve may not look like this one, so be careful to remember the sequence of parts when you remove the valve to clean it.

perfectly flat, which means that the entire unit has to be hauled off to a machine shop or replaced.

Spout leaks—Any or all of the possible causes of stem leaking can also cause the faucet to drip. Disassemble and examine as suggested.

SINK SPRAYS

Sink sprays stop working or work poorly for any number of reasons. The most common cause is clogged aerators on either the nozzle on the end of the flexible hose or the faucet's common spout. When this occurs proper pressure changes do not develop in the valve chamber when you push the spray button, and the diverter valve does not operate. Correct this problem by carefully removing the aerators and washing them free of debris. Be certain to replace all the parts in the correct order or the device will not work.

Another frequent cause of poor spray performance is a clogged or defective diverter valve. You will find this valve below the spout. Remove the spout to get at it. Open one faucet gently and let the water push the valve up and out. Clean the valve chamber by letting more water flow out. Then clean the valve itself by washing it and scraping it free of any accumulated mineral deposits. If cleaning the aerators and the valve doesn't produce proper spray operation, replace the diverter valve.

CHAPTER NINE

Fixing Pop-up and Drain Valves

The metal valve that prevents water from leaving your lavatory bowl is usually a pop-up valve. You can see the valve "pop" upward as you operate the control lever or knob. The metal valve that prevents water from leaving your bathtub may be a pop-up valve or a drain valve. If a metal plug is visible over the drain hole, it is a pop-up valve. If no plug is visible you have a drain valve.

LAVATORY POP-UP VALVES

Old-fashioned lavatories have a vertical handle behind the central spout that operates the pop-up valve. Moving this handle up and down acts to lift and lower the metal plug (valve) in the center of the drainpipe. On newer lavatories the same valve is operated by rotating a short handle mounted on a ring directly behind the spout.

The lower end of the aforementioned handle is connected to a pivoted horizontal bar. The other end of the bar is connected to or supports the pop-up valve. The valve itself is a casting or tube having an upper end that is larger than the drain opening. Thus when the tube (valve) is raised the drain opening is partially uncovered and water can leave the lavatory basin. When the valve is lowered it seals the drain opening and water remains in the basin.

The same action is used on the newer lavatories, but instead of operating the valve by the up and down motion of the valve handle, the handle is rotated and its motion is transferred to the same vertical rod. The two parts are loosely connected.

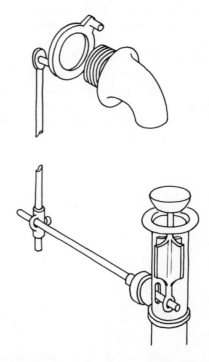

113. Basic arrangement of lavatory pop-up valve and control mechanism. On some models, a vertical control handle is used in place of the rotating handle positioned behind the spout.

Vertical handle stuck—This is caused by lack of use and possibly by the prolonged action of soapy water. Try a little liquid wrench followed by thin motor oil on the vertical shaft. If that doesn't loosen it up sufficiently, disassemble the rod, remove it and rub it down with fine emery cloth or steel wool. Lubricate and reassemble.

Rotating handle stuck—This is usually due to the same causes—corrosion and/or an accumulation of hard soap. Try some liquid wrench and thin oil. To disassemble, wrap some adhesive tape around the spout and use a Stillson wrench to remove it. Tape prevents spout from being chewed up by wrench. Again, rub corrosion off with emery cloth or steel wool. Lubricate and reassemble.

Valve doesn't move enough—This is a common problem and is due to poor initial design and wear. Even when new, the valve rarely moves more than ⅜ inch from fully open to fully closed. Therefore, when parts wear at their joints just a little the motion of the valve is seriously affected.

Try to take up the wear at the joints (play) by readjusting the various parts, by bending the rods a little and by turning the valve and rods a little so that contact is made between unworn surfaces rather than worn surfaces. If that doesn't produce sufficient valve swing, replace the parts one by

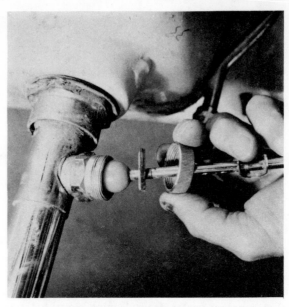

115. Close-up of pivot rod, packing ring or washer, and pivot ball.

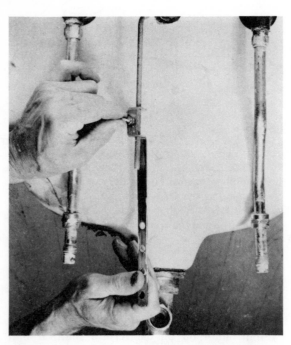

114. To adjust height of pop-up valve in bowl in relation to the position of the control, the distance between the control and the pivot rod is varied.

one. Sometimes just replacing the most worn part will give you several more years of service.

Valve cannot be fully opened or closed—This is caused by maladjustment between the levers. Usually the horizontal rod or bar can be adjusted in relation to the vertical rod. Try moving the clip or clamp in one direction and then the other until you get the swing you need.

Valve won't hold water—If the valve is in its fully closed position and still leaks, the trouble may be a bent or worn or dented valve or drainpipe, or a worn or torn valve gasket. Look for this gasket just under the valve top. Replace it if it is damaged.

Water drips from side of drainpipe—This is due to a leak between the pivoting horizontal rod and the drainpipe fitting. The latter is usually a ball-and-socket joint sealed by string or O-ring packing. Try tightening first. If it still leaks, disassemble the joint and replace the packing.

TUB POP-UP VALVES

These work on the same principle as the lavatory pop-up valves. The operating mechanism is within the overflow tube, however. This means that you have to remove the lever control from the overflow tube to make adjustments. The most common tub pop-up valve design is the snake pop-up.

Snake pop-ups—This valve consists of a metal plug that fits into the drain opening in the bottom of the tub. The underside of the plug is attached to a series of jointed metal rods about 8 inches long overall. The end of the snake points upward, into the overflow tube. The control may consist of a short lever pointing into the tub, or a knob that is rotated. In either case, actuating the rod or knob moves a rod within the overflow tube. The lower end of the rod touches the end of the snake. The point to bear in mind here is that the snake and its plug work by weight alone. To open the valve, the inner rod is moved down, pushing the snake up and out. To close it, the rod is lifted and it is the weight of the snake alone that closes the valve. Therefore, hair and other gunk that collect on the jointed rod will prevent the valve from properly closing.

To remove the snake, just pull it out. To replace it, push it back in again. This is not always easy to do, but if you try long enough, twisting and turning the snake's tail, you will eventually find the proper slot and the snake will fully enter the drainpipe.

DRAIN VALVES

This term is most often used for a lavatory valve or a tub valve that is positioned within the drainpipe and out of sight.

Pivoted lever—This design has a' short control lever that pokes through a metal cover plate into the tube. The cover plate and lever are usually positioned directly beneath the tub's spout. Moving the lever up and down opens and closes the drainpipe by virtue of a short brass plug that enters a reduced-diameter section of the drainpipe. The brass plug, the connecting rod and spring and the inner end of the control lever are housed within the overflow tube. The overflow tube runs from the drainpipe to an opening in the tub just behind the cover plate.

When the valve will not remain in the open position—that is to say, the control lever will not remain down—the trouble is almost always due to wear at the pivot point. Replace the lever and plate. These parts wear very quickly because they are made of white metal, which is soft.

When the valve fails to shut off the flow of water out of the tub the trouble is almost always due to debris at the valve. Lift the plug by lowering the control lever and run a snake through the

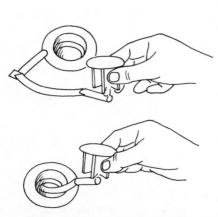

116. Snake pop-up valve. It's not difficult to remove: just lift it up and out. Getting it back in again may involve a lot of twisting and turning, but it will go in; just keep trying.

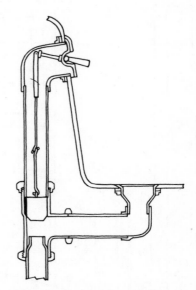

117. One type of drain valve. Malfunction usually results from wear at the control-lever support.

drain. Then remove the cover plate and lift the mechanism out. Now poke a stiff wire down into the valve opening to remove whatever debris you find there.

On rare occasions the valve will not open fully or close fully. This is due to the connecting rod length being incorrect for the plug and the control lever. Remove the cover plate and lift the connecting rod out. There is usually a point for making a length change somewhere on the rod. If the valve doesn't fully open, shorten the rod. If it doesn't fully close, lengthen the rod.

Lift tube—The oldest and still best all-around design consists of two concentric tubes standing free of the tub. The bottom of the outer tube connects to the drainpipe. The bottom end of the inner tube fits snugly into the drainpipe when it is lowered and in its closed position. To permit the tub to drain, the inner tube is lifted.

Failure of this type of drain valve to completely shut off the flow of water out of the tub is usually due to debris collecting at the drainpipe where the inner tube makes its seal: the valve.

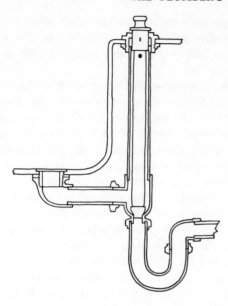

118. Details of a lift-tube drain valve. Lift the inner tube up and out to gain access to the trap. A length of stiff wire with a hook at its end will help you remove obstructions from the drain.

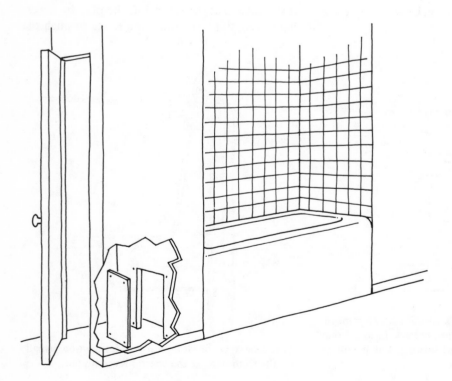

119. Plumbing codes require that access be provided to the underside of the tub's drainpipe. Look for a trapdoor in a linen closet that backs up against the tub.

place to try next. Access to this point in the plumbing system is required by all plumbing Remove the inner tube and poke the debris free with the end of a snake or a stiff piece of wire. Also try pouring a mixture of hot water and a chemical drain opener down the tube, See Chapter Eleven for data on chemical drain openers and their use.

EXTERNAL ACCESS

On a few tubs the overflow pipe or the associated piping has a removable plate, bolted and sealed in place with a gasket. When repairs and adjustments to the pop-up or drain valve mechanism cannot be made from the tub, this is the codes, so look for a trap door in an adjoining closet. If there is no trap door, do not cut a hole in a wall or ceiling: the chance of finding a removable plate on the drainpipe is very slim in such a case, and if you start disassembling the drain- and overflow pipes on an old house you are looking for trouble.

Instead, take twenty-five cents, go to the five-and-dime store, purchase a reliable rubber plug and use that. Mess with an old, partially corroded drainpipe and you will end up replacing everything in sight.

CHAPTER TEN

Fixing Leaking Pipes and Joints

Your first step in repairing a leak in a pipe or in a pipe joint is to determine whether or not the pipe and joint carries water or soil. Whereas water lines are under pressure and require "solid" repairs, soil lines are not, and in an emergency soil lines can almost be repaired with chewing gum.

The determination is easily made. In a home, pipes larger than 1¼-inch outside diameter are almost always soil. Leaks that squirt or drip continuously are always water (pressure) leaks.

SCREW-THREAD JOINT LEAKS

With a union nearby—Screw-thread leaks can sometimes be stopped by simply tightening the pipe into its fitting. When there is a union nearby in the line, tightening is easily done. The large nut on the union is backed off a turn or so. Now the pipe is tightened into its fitting (in some cases you may find it easier to turn the fitting). The union is then closed. In some instances you can do this without draining the pipeline as very little water will come out of the union if it is just loosened. In other instances you will have to close the line and drain it first.

With no union nearby—This usually doesn't work, but it is worth a try as it is easy to do.

With full pressure on the line, slowly and carefully tighten the pipe into the fitting at the joint that is leaking. Of course, when you do this you will simultaneously loosen the other end of the pipe in its fitting. No matter. If a joint has been properly made and doped it sometimes can be

backed off a fraction of a turn without producing a leak.

Wait it out—If there is no union in the line, and if you cannot stop the leak at one end of the pipe without opening a second leak at the other end of the same pipe, you can opt for waiting it out.

To do this, turn the pipe so that you have two small leaks, one at each end. Place buckets under the leaks and just wait. With a little luck, lots of time and slightly hard (lime filled) water the leaks will seal themselves closed. This approach works best with hot water, but it can take months.

Tightening doesn't help—In some cases, even when you have a union handy and can tighten up as much as you wish on the offending joint, it will continue to leak. This is because the threads on the pipe or fitting were not properly made, or the joint was made up dry, or the joint was overtightened and then backed off. In any case, you have to take the joint apart. If you find it dry, try dope and see if that holds. If that doesn't do it, try a new length of pipe. If the new pipe and dope and moderate tightening pressure still do not stop the leak, the fitting's thread is defective and must be replaced.

Emergency threaded-joint repairs—When you can't or don't want to take a threaded joint apart, there are ways of stopping the leak that sometimes work. A lot depends on how much pressure there is in the line.

The line is shut down and drained. The faucet at the end of the line is opened. Then the joint is

120. There are several special plumbing cements made for the express purpose of sealing holes in high-pressure pipelines. Some of them even purport to work with the pipe wet, but none will work with pressure in the line. For best results, however, dry and scuff the pipe surface before applying the cement.

heated to make it completely dry. The external surface of the joint is scrubbed clean with steel wool. Epoxy cement or special plumbing cement, made for the purpose, is generously smeared over the end of the fitting and the exposed thread. Make the cement about ¼-inch thick; carry it up onto the fitting for at least an inch and an equal distance down the pipe. Let the cement dry thoroughly. Then give it a try.

If you can unscrew the joint but can't or don't want to replace the pipe and/or fitting: scrub the pipe threads and the fitting threads clean with steel wool. Smear epoxy cement on the thread and reassemble the joint, making it little more than hand tight. Let the cement dry. That usually does it, but the joint can never be taken apart by normal means again.

UNION LEAKS

When the large nut on the union is tightened it presses the two machined faces together and a watertight joint is formed, if the faces are perfectly smooth, clean and fairly parallel before the nut is tightened. If these conditions are not met the union will leak.

Loose nut—When you encounter a leaking union the first and easiest thing to try is tighten-

ing the large nut. You need two wrenches to do this, one to hold the end of the union and the other to hold the large nut. Obviously, if you turn the large nut the wrong way you will loosen the joint. To make certain that you do not, examine the large nut. Note that one end is internally threaded and that the other has a smaller opening, which is not threaded. As you face the end with the smaller hole, turn the nut clockwise to tighten it.

Defective surfaces—If tightening doesn't stop the leak, close the line and drain it. Then disassemble the union by backing the large nut off. Spread the facing surfaces. This is not always easy to do. Sometimes the attached pipe can be "sprung" (pressed to one side). If not, you will have to unscrew one half of the union or the associated section of pipe. If you cannot see the union faces you cannot tell what is wrong.

If the union has been leaking for a long time there may be a rusty groove across one face, or both surfaces may be grooved and corroded. Depending on how badly grooved the face(s) may be and how much time you want to spend, you can decide whether to attempt to repair the old union or to install a new one.

To fix the old union, burnish the faces clean with steel wool. Dry them and apply a layer of epoxy cement in the grooves and corrosion pits. When the cement is dry and hard, use fine sandpaper on a smooth board to sand the surface flat. Wipe the surface clean and apply a layer of pipe dope on the mating faces. Reassemble the union and hope for the best.

As an alternative you can clean the faces with steel wool and apply nonhardening, gasket-form-

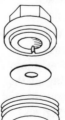

121. When a groove has been worn across the face of a union by a persistent, long-term leak, use a layer of automotive gasket cement or a rubber gasket to seal the joint between the two halves of the union.

ing, automotive-type cement to the mating surfaces. Assemble the union and check the seal. If you use epoxy cement on the mating surfaces you will not be able to ever disassemble the union again.

Still another alternative consists of cutting a gasket from automotive gasket sheet and fitting it between the mating surfaces. This remedy almost always works, but you must have sufficient clearance to position the gasket without tearing it.

CRACKED SCREW-FITTINGS

A hard frost will crack water-filled pipe and fittings. Generally the crack is not at the threads, but in the body of the fitting. Such cracks can sometimes be repaired with epoxy cement or with one of the liquid metals sold by plumbing supply shops. The pipe has to be drained thoroughly. Use a torch on the fitting to save time. Then smear the cement into the crack. All these cements set faster and harder when they are warm, so use an electric heating pad or a lightbulb when the temperature is low.

For the best possible repair, disassemble the joint and press the cement that enters the fitting through the crack, back against the inner fitting wall. This expands the cement into the crack, permitting better adhesion, and also reduces the space otherwise occupied by the cement inside the fitting.

SWEATED-JOINT LEAKS

Sweated (soldered) joints that are properly made never leak. Thus, if a sweated joint leaks it was improperly made, which means it must be remade. The reason the joint did not leak for months or even years is that the leak was plugged by flux or it was sealed by pressure. Eventually the flux dissolves; eventually the house settles a little and the pressure is relieved and the joint leaks.

Open the tube—The first step is always to close the line, drain the water and open one end of the tube to the air. The latter can be done by opening any faucet connected to the tube. If the joint is next to a valve, disassemble the valve and remove its stem and attached washer.

Reheating the joint—Sometimes you can close a leaking sweated joint by reheating. It doesn't work very often, but as it is easy it is worth a try. Heat the joint as explained in Chapter Five, but this time use acid-core solder wire in place of solid solder wire. The thinking here is that the acid will clean up any internal dirty spots that weren't tinned the first time: these may have caused the leak.

Remaking the joint—Assuming that simply reheating did no good, close down the line again, drain it and open it to the air. Reheat the joint and pull it apart when the solder has melted. Examine the surface of the tube end you have removed. Examine the inside of the fitting. Rub clean any dirty spots and bare copper spots with steel wool or sandpaper. Next tin the bare spots. Do this by applying flux and heating the metal a distance away from the flux until the metal is hot enough to melt solder. Then the tip of the solder wire is touched to the bare spots, covering them with a thin coat of solder. When the metal is cool, reassemble the joint, heating it as you do to help you get the tube inside the fitting. Heat it some more to make certain that the solder is truly melted. Feed the joint a little solder and it is done. Withdraw the flame and let it cool.

Alternative to soldering—A leaking sweated joint can sometimes be sealed without soldering by the use of epoxy. The joint must be perfectly dry and warm so that the cement dries quickly. The cement is applied to the joint and forced in, if you can see the hole. Additional cement is applied as a layer around the joint and up and down the pipe and fitting for about 2 inches. If you can get cement into the leak, if the surfaces are perfectly clean and dry, epoxy will hold. But if it doesn't and you have to go to solder you will have a difficult time of it removing all the epoxy; you cannot solder when the metal is covered with epoxy.

PLASTIC-PIPE JOINT LEAKS

Like sweated joints, correctly made plastic joints do not leak. When they do the fault lies in their making. Unfortunately, plastic joints cannot be disassembled. Once they are made they are made for the life of the system.

Sealing leaking joints—One trick you might try consists of closing the pipe, draining it and drying it—plastic solvent will not adhere if water is present. Then force some solvent into the leak between the fitting and the pipe. Give it 10 minutes to harden and then try the water. If that fails the only certain solution is to replace the joint.

Replacing a plastic joint—This is not as difficult as it might sound, but it is time-consuming and costly in material. The joint is cut out with a hacksaw. A new fitting plus two short pieces of pipe and two slip fittings are added. The short pipes are cemented to the fitting and the sleeves are used to join the short pieces of pipe to the rest of the pipe.

THICK-WALL PIPE LEAKS

Thick-wall pipe such as standard galvanized, black, brass and plastic rarely crack along their length, except when burst by frost. However, they do crack, and when they do it is generally much easier and less expensive to repair the crack than to replace the section of pipe, especially when the crack is in the middle of a run and there is no union nearby.

Cementing the crack—The pipe is drained and dried. Epoxy cement or one of the liquid metals made for the purpose is forced into the crack. When the cement is dry the crack will be sealed, if you are lucky.

Soldering the crack—Cracks in copper and brass pipe can be soldered closed. Use the same tools and techniques used for soldering copper tubing that are discussed in Chapter Five. Drain and dry the pipe. Clean the crack. The metal edges must be bright—free of all grime and oxide —or the solder will not bond. Then the crack is fluxed and the nearby pipe is heated. When the pipe is hot enough to melt solder, the joint is filled and sealed.

Cracks in plastic pipe—Drain and dry the pipe. Force plastic solvent cement into the crack and push together the sides of the pipe with a clamp; hold them in position until the cement sets.

Clamp repairs—Cracks in pipes can sometimes be permanently closed with clamps of all sorts, including clamps made especially for the purpose.

The commercial pipe-leak clamp consists of two sections of curved metal plus a rubber sheet. The rubber is placed over the leak and the two halves of the clamp are bolted together and tightened against the leak. The advantage of this kind of repair is that it is fast. It can be made with the water running, and the clamp and rubber will last for years.

In an emergency C-clamps and automotive radiator hose clamps can be used. A clamp can even be rigged from a block of wood and baling wire. To use the C-clamp a block of wood is placed over the crack and then the clamp is used to hold the wood firmly in place. The strap-type automotive clamps are used with a split rubber sleeve, which can be a section of rubber hose or a thick sheet of rubber. The flexible material is positioned over the leak and the strap is used to hold it in place. If the crack is long, use two strap clamps and a strip of metal to back up the rubber where it rests on the crack.

If you lack clamps, place a block of soft wood over the crack and tie it loosely in place with half a dozen turns of galvanized wire. Then use a pair of pliers to twist the wire to apply tension. If you wrap the wire in coils you can develop a lot of

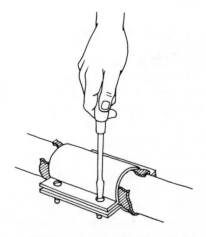

122. A commercial pipe-leak clamp makes a permanent seal and will keep a pipe leak closed for years. Clamps are available in a number of sizes to fit various-diameter pipes.

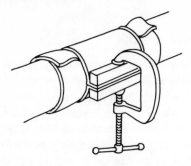

123. You can make use of a C-clamp to stop a pipe leak if you take the trouble to bend a C-shaped piece of metal around the pipe. A layer of rubber is placed between the metal and the pipe, and two strips of wood keep the clamp fast onto the metal.

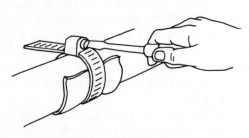

124. A strap-type automotive clamp and a section of rubber tubing will quickly stop a small leak. For larger cracks or holes, use a pair of clamps and a curved piece of metal to back up the rubber.

pressure when you twist it. Also, if you cut a groove to fit the pipe in the underside of the block of wood before you wire it in place, a better seal will be secured.

REPLACING PIPE

Shut the line down and drain it. If there is a union, take it apart and unscrew the defective section of pipe. Replace it with a new section. That was easy, but it doesn't always happen that way. Most of the time there is no nearby union. In such cases the following approach always works.

Cut the pipe at an angle with a hacksaw. Make the cut close to one fitting, rather than in the center. Bend the longer of the two pieces of pipe and unscrew it with the help of a Stillson wrench. Unscrew the short section. Replace the short section with a new short section—a nipple, to be exact. Connect a union to the nipple. Take the old long section, if it isn't defective, and hold it in place between the new union and the old fitting. Mark it, cut it and thread it. Then take the union apart. Connect it to the new pipe. Insert the other end of the pipe in the old fitting. Screw it home and reassemble the union.

COPPER-TUBING LEAKS

Copper tubing is easily damaged. A nail will go through it without pause. A misdirected wood chisel will open a gash. Fortunately, holes in copper tubing are easily and permanently repaired by soldering.

Nail holes—Drain and open the line. Rub the area around the hole bright with steel wool or sandpaper. Flux and solder it closed. In doing so, keep the torch near the hole, but not directed onto the flux. Feed the tip of the solder wire against the area around the hole until it has been tinned. Then fill the hole and withdraw the torch immediately—you don't want the solder to fall into the hole. Water can be immediately admitted to the pipeline.

Cuts—Large nail holes and cuts in the tube are treated the same way. The area is cleaned and fluxed and tinned. However, do not depend on solder alone to seal a hole that is more than $\frac{1}{16}$-inch across. Use a patch. Make it out of a small piece of copper; a section of tubing will do. Bend it so that it fits smoothly over the hole and cut it to extend perhaps ½-inch beyond the edges of the hole. Then tin its underside and place it over the hole. Now heat both the tube and the patch and feed solder into the joint when it comes up to temperature. As both surfaces are tinned, there is no need for flux.

Properly soldered holes and patches can last the life of the copper tubing. There is no need to worry about them ever opening.

REPLACING COPPER TUBING

Should a section of your tubing ever become so mashed that it cannot be patched, it is a simple matter to replace it—far easier than to replace a section of thick-wall pipe.

Use your pipe cutter to cut out the defective section. If one end is close to a fitting, unsolder that end from the fitting after severing the tube. Cut a new section of tubing ¼ inch shorter than the exact length of the tube you remove. Pass one slip fitting over each end of the new tube. Position the new tube in line. Solder the two slip fittings in place, again as per instructions in Chapter Five. When the joints are cool the job is done.

PLASTIC-PIPE LEAKS

Like its metal counterparts, plastic pipe does not leak by itself. It has to be damaged by frost or stray nails, chisels and the like. But unlike copper or galvanized pipe, damaged plastic pipe is easily repaired.

Using patches—Close and drain the line. Wipe the damaged area bone dry and then remove its shine with sandpaper. Cut a plastic slip fitting lengthwise in half. Remove the shine on the inner section with sandpaper. Give both sanded surfaces a coating of plastic solvent cement. Place the patch over the hole and slide the patch back and forth just once to spread the cement evenly. Press the patch in place with your fingers. Wait a few minutes and the job is done. However, it is wise to give it an hour or so to harden, especially if the temperature is much below 60° F.

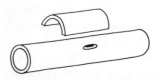

125. The easy way to close a hole in a plastic pipe is to cover it with a patch. Empty and dry the pipe; then prepare the patch and the surface just as you would if you were joining plastic pipe to a fitting. Hold the patch firmly in place for a few minutes to make certain of a good weld.

SOIL-PIPE LEAKS

All the repairs suggested for pressure-pipe leak repairs are applicable to soil-pipe (low-pressure) leaks. Not only are they applicable, they are recommended. Although you can patch most soil pipes with a little adhesive tape, it is inadvisable to do so except in an emergency. Soil contains an abundance of pathological organisms, so even a little seepage can be dangerous.

Caulked-lead-joint leaks—Joints between sections of cast-iron pipe and between cast-iron pipe and fittings are usually made by lead caulking. Very simply, **oakum** (a kind of rope) is pounded into the space between the **socket** (expanded pipe end) and the unaltered pipe end to which it is to be joined. This is followed by molten lead, which is permitted to cool, then also pounded and expanded to lock it and the oakum in place. Untouched, lead-caulked joints will hold for almost forever. Should the building settle—this is not uncommon—causing the lead joint to bend a little, it will leak.

Small, lead-caulked leaks can be stopped by driving the oakum back into its socket and expanding the lead to hold it there. This is easily done with a blunt-ended rod or bar and a hammer. Tap lightly until the leaking stops. If this fails, dry and seal the joint with a little epoxy cement or liquid metal.

To dry the drainpipe, shut the main valve off the night before, after flushing all the toilets. By morning the pipe should be dry. If not, heat it a bit with the torch. If the crack area is rusted or

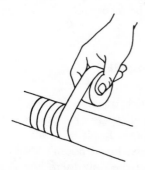

126. Soil pipe leaks can be temporarily closed with adhesive tape or even masking tape.

mucky, clean it bright before applying the cement.

THIN-WALL BRASS PIPE LEAKS

Thin-wall brass pipe is used almost exclusively for sink, lavatory, shower and tub drain connections, that is to say, from the fixture to the in-wall galvanized, plastic or cast-iron drainpipe. When the thin-wall pipe is out of sight it is usually bare. When it is exposed it is plated and polished, but it is still the same thin pipe. The only difference from a repair point of view is that the bare pipe will show the ravages of corrosion, whereas the plated pipe will look perfect until it falls apart.

Joint leaks—When a thin-wall brass pipe joint is knocked askew and then straightened, it sometimes leaks because the pipe end is bent and doesn't make a perfect seal against the flexible compression ring. Take the joint apart. Replace the pipe section if its end is bent or dented. Make the replacement section long enough to prevent the joint from being damaged. To keep the joint from ever being knocked askew, straighten the original pipe section. Remove the ring and compression nut. Clean, flux and solder the pipe to its fitting. See Chapter Five for soldering information.

Thin-wall brass pipe joint leaks are also caused by loose compression nuts and rotted compression rings. First try tightening, bearing in mind that when these joints are in good condition they will hold water when only hand tight. Next, if the nut is tight or tightening doesn't help, take the joint apart and replace the compression ring. If that doesn't do it, look for a leak in the next higher joint. Very often the water seeps down the pipe so evenly you are not aware of its actual source.

Compression-nut corrosion is caused by drainpipe sweat, (for which the only practical cure is to tape the joint and the pipe to prevent moisture condensation), by iron wire and by leakage.

Iron or galvanized wire is sometimes wrapped around the brass drainpipe as a support for something or other—to hold a pot, for example. The galvanic action that develops when the pipe

127. Thin-wall brass pipe is easily cut with a hacksaw and a simple guiding jig.

and wire are moist quickly eats a hole in the copper.

Small, hardly noticeable leaks at the joint will in time corrode the compression nut. These nuts are often made of white metal or some other foul alloy that is galvanically dissimilar to the brass. When moisture is present they both corrode rapidly.

Pipe leaks—With time and especially when assisted by a nail or hair clip or pin, thin-wall brass pipe corrodes through from the inside out. You usually discover this fact when you attempt to replace a compression nut and the pipe just crumbles in your hand. Should your pipe leak before the compression nut has to be tightened or replaced, place a pot under the drip and wait until the plumbing supply shop opens. Frequently the pipe will collapse if you try to tape the hole closed.

Thin-wall brass pipe is cut with a hacksaw. Use a fine-toothed blade and hold the pipe itself in a miter box. You can make a miter box from scrap lumber.

LEAD-PIPE LEAKS

Today, lead pipe is only used for the toilet bend, which is the elbow that connects the bottom of the toilet to the drainpipe. (If your home has a lead pipe leading from the water service pipe to the water meter, remove it and replace it with anything else. Lead water pipe causes lead poisoning.)

Although lead is well-nigh infinite in life-span it does crack comparatively quickly: its grains separate. Lead pipe cracks can usually be satisfactorily sealed for many years with a plumber's poultice, which is a cloth and plaster of Paris bandage, much like the cast used on a broken leg.

To make and apply the poultice, wipe the lead pipe clean and apply a thin layer of wet plaster all around the pipe for a distance of 3 or 4 inches above and below the ends of the cracks. Wrap one turn of clean cloth around the plaster and pipe. Apply another layer of plaster and wrap the cloth some more. After 4 or 5 layers of plaster and cloth, finish up with a layer of plaster. Let it all dry hard and the job is done.

CHAPTER ELEVEN

Opening Plugged Fixtures and Drains

OPENING PLUGGED TOILETS

Every toilet has a built-in trap, an internal loop in the ceramic pipe leading from the toilet bowl to the soil pipe underneath. If the work was done according to the plumbing code, the diameter of the soil pipe forming the trap is the same or smaller than the diameter of the soil pipe following. This is done to prevent obstructions from entering the soil pipe where they would be more difficult to remove than from the trap, and also to speed the flow of soil out of the trap. Therefore, whatever is preventing the flow of water and soil out of the toilet bowl is most likely in the toilet trap. Clearing a plugged toilet, then, is usually a matter of either pulling the obstruction out of the trap or pushing it on ahead into the larger soil pipe.

Prevention—Toilets do not plug up in ordinary use. Obviously they will become plugged if someone drops something into the bowl that is too firm to dissolve and too large to pass through. To preclude this possibility it is wise to keep the bathroom clear of all small objects that might fall into the bowl, and locked to small children who like to throw things into the bowl.

Another, less obvious, cause of plugging is the use of paper napkins in place of toilet paper. Napkins are formulated not to dissolve in water, whereas toilet paper is made to dissolve. While it may not appear likely, a handful of soft paper napkins can plug a toilet as quickly as a child's toy.

Warning signs—When something large lodges in the toilet trap, stoppage is almost immediate. When something small does the same, stoppage may be no more than partial. Soil flow is slower, but still may be complete. The tip-off is that the bowl takes longer to empty and once in a while it doesn't empty completely, but needs a second flushing. Since this may also be due to a lack of sufficient water in the tank (see Chapter Seven), the water level should be checked first. If there is plenty of water in the tank and if emptying becomes increasingly sluggish, toilet paper is building up around a small obstruction and will in time cause complete stoppage. Obviously it is best not to wait for that to happen.

Don't risk a flood—When the toilet is reported as being plugged, do not try it by flushing it. Instead, remove the tank cover, raise the stopper ball and let a little water enter the bowl. Should the water level in the bowl rise dangerously high, push the ball down to stop the flow.

Call a "friend"—The simplest and least mucky tool you can use to open a plugged toilet, be the cause a large or a small obstruction, is the plumber's friend or plunger. Two types are sold. One is a simple semi-hemisphere of rubber attached to a handle. The other has a sort of tube extension from the bottom of the rubber semi-hemisphere. This is the far better type, as it develops more pressure. Purchase the heaviest, stiffest rubber plunger you can find. The stiffer it is the more effective it is.

To use with a toilet bowl, extend the tube and insert it into the opening in the bottom of the bowl. Let sufficient water into the bowl by means of the hand-operated stopper to cover the plunger.

128. Using a force pump, sometimes called a plumber's friend. Make certain that there is sufficient water in the bowl to cover the bottom of the pump; pump slowly and steadily.

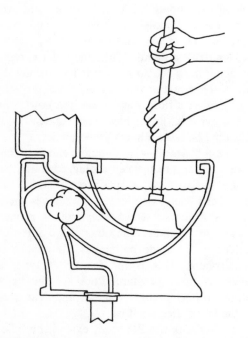

129. The force pump operates by driving the obstruction out of the built-in toilet trap. As the obstruction moves, the water level in the bowl will drop; add more water and keep pumping.

Press the bottom of the plunger firmly against the bottom of the bowl and work the handle up and down. Make the down stroke as rapid as you can, the purpose being to drive the water against the obstruction. The speed of the up stroke is unimportant.

Sometimes a couple of pumps opens the line—sometimes not. Don't lose heart and give up too quickly. It sometimes takes ten or fifteen minutes to get results. If the water level in the bowl moves down, you are gaining. Let a little more water in and continue.

Pumping all the water out of the bowl doesn't always indicate clearance, so carefully let more water into the bowl before you replace the tank cover and consider the job done.

Whether or not the plunger will clear the trap depends on the nature and size of the obstruction. If it is a collection of soft material, chances are that you can pump it down and out if you work at it long enough—each plunger downstroke moves the obstructions a mite farther down the pipe. If it is soft material around something small but hard—a spoon, for example—it is possible to push the soft stuff past the spoon without moving the spoon very much. In such cases the toilet will plug up at regular intervals and will continue to do so until you remove the small but firmly positioned obstruction.

Using a closet auger—When the plunger fails to open the line at all, or when regular, repeated stoppage indicates that a portion of the obstruction remains, move on to the closet auger. This tool is a short snake attached to a handle and enclosed in a curved metal tube. In selecting an auger, it is again advisable to purchase the strongest you can find. The soft, thin-cable augers are a waste of time and money.

To use, pull back on the handle to bring the end of the snake close to the tube end. Direct the tube up and into the end of the bowl where the large hole that leads to the trap is. Then hold the tube firmly in one hand while you push and turn the handle. Doing so drives the tip of the snake forward. If you can't turn the handle in one direction, try the other or work it back and forth, always seeking to push the snake's tip forward. When you feel you have reached the obstruction, try turning the snake and pulling it out, hopefully

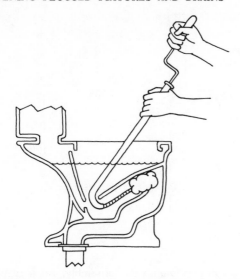

130. Using a closet auger to remove an obstruction in a toilet. Note that on this particular toilet bowl outflow is toward the front. On other bowls, such as the one pictured with the force pump, the flow is rearward. Be certain to guide the auger in the right direction; otherwise there is a chance of breaking the lip of the jet guide (at rear of bowl in this toilet).

of the snake coiled in your hand so that you can easily rotate the snake as you urge it on. You can use the snake to push the obstruction ahead and into the soil pipe, or you can uncoil the end of the snake to make a hook with which you can pull the obstruction up and out.

Removing the toilet—When all else fails the toilet has to be temporarily removed. Doing so exposes the trap and the soil pipe opening. Although the removal and replacement of a toilet is a couple of hours' work for even an experienced plumber, it isn't very difficult, and you can be almost certain that you will clear the stoppage if you do this. Toilet removal and replacement is discussed in Chapter Twelve. However, before you attack the toilet make certain that the stoppage isn't in the following soil pipe and that you cannot get to the blockage any other way except by removing the toilet. Drain and soil pipe blockages are discussed a few paragraphs on.

hooking on to the obstruction and pulling it along. If this doesn't work, try pushing the obstruction forward. With a little luck, lots of patience and considerable effort you probably will remove the blockage.

Using the snake—The closet auger will not do it every time. Sometimes the obstruction is too tightly lodged in the trap and sometimes it is beyond reach of the auger. When this is the condition, try the snake. Snakes aren't tried until the plunger and auger fail, because snake work is messy; you have to place your hands inside the bowl and there is a possibility that the snake will chip the bowl.

Once again you are advised to purchase the heavy snake made of curled wire rather than the thin type or the flat-strip type of snake.

Guide the end of the snake up inside the bowl and then down over the lip of the trap. This is where the bowl is often chipped, so be careful. A chip doesn't ruin the bowl, but it causes a permanent stain where the glaze is removed. Keep most

131. When the obstruction is in the soil pipe following the toilet, the toilet has to be removed. When this is done you can guide a large snake down the toilet bend as shown.

Using chemicals in toilets—No mention has been made of the use of chemicals for clearing toilet stoppages. The reason is a good one: Chemicals are almost useless for toilet stoppages. The toilet trap has an internal diameter of 3 inches; the soil pipe following has an internal diameter of 3 or 4 inches. This makes for a lot of water, and unless you pour chemicals down by the gallon, they will have little effect. Even if you do use large quantities, chemicals will not remove a jammed spoon or a stuck plastic toy. The use of various chemicals for unplugging sinks and waste pipes is discussed a few paragraphs on.

OPENING PLUGGED SINKS, LAVATORIES AND TUBS

In addition to becoming plugged by foreign objects, sinks, lavatories and tubs become plugged up in the course of normal use.

Warning signs—Like toilets, sinks, lavatories and tubs rarely stop emptying all at once. Generally, discharge gradually slows down, warning of impending complete stoppage. The warning should be heeded, since it is much more difficult to open a completely blocked fixture drain than it is to open one that isn't completely blocked.

Foreign objects—Lavatories are plagued by an accumulation of hairpins, toothpaste caps and toothbrushes. These objects do not of themselves stop the flow of soil, but they do hasten the accumulation of hair, lint and grease.

Kitchen sinks are often troubled by table knives. They slip vertically down into the drainpipe. One or two knives do not slow the flow of waste very much, but they speed the collection of kitchen grease, which eventually does stop soil flow.

Bathtub drains are the repositories of buttons, hairpins, nail files, hair, grease and solidified bubble bath compounds. Washtub drains suffer a collection of coins, buttons, safety pins, lint and hardened soap.

Remove or open the pop-up valve—The first step to unplugging any of the fixtures mentioned consists of opening or removing the pop-up valve and exposing the drain opening.

Those pop-up valves that are visible and that work by stoppering the top of the drainpipe

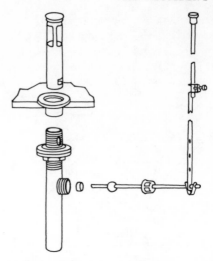

132. Typical pop-up valve assembly. Some valves can be removed by simply lifting them straight up; others must be turned or tilted first.

where it enters the fixture can usually be removed by simply lifting them up. Some of these, however, need to be given a fraction of a turn to the right or left before they can be lifted up and out. Still others have to be raised to their open position by means of the control lever. Then the lower end of the valve is swung free of the operating bar inside the drain.

The drain valves that operate within the drainpipe and that are out of sight are merely opened all the way and left in that position until the job is done.

Using the plunger—The plunger is fairly effective in opening plugged drains that are not connected to an overflow tube. Sinks, washtubs and old lavatories usually do not have overflow tubes. New lavatories and bathtubs usually do. When there is no overflow tube full plunger pressure can be exerted against the obstruction. When there is an overflow tube, pushing smartly down on the plunger merely pushes upward on the water in the overflow tube. To some extent this pressure loss can be corrected by firmly pressing a wet rag against the overflow opening in the side of the lavatory or tub (it is usually under the pop-up valve lever).

To use the plunger, fill the bowl or tub with a couple of inches of water. Fold the tubular extension back into the semi-hemisphere. Press the bottom of the device firmly over the hole and work the handle up and down. Make the downstrokes more rapid than the upstrokes in order to drive the water down into the drain. If the water level drops you are gaining. If the water level doesn't drop give it at least 10 or 15 minutes before giving up. Few drains can be opened with a dozen strokes.

USING CHEMICALS

If the plunger fails, but there is some water movement, you can use any of the drain-opening chemicals made for the job. When there is no flow it is inadvisable to use any chemical, no matter what is said and illustrated on TV. Very simply, should you pour a load of powerful chemicals into a stuffed sink or tub and should the chemicals fail to open the drain, you will have a tubful of dangerous chemicals to contend with in addition to a stuffed drainpipe.

Choice of chemicals—Two types of chemicals are used: alkali and acid. The alkali include lye, Drano, "99" and others. Lye, which is caustic soda, is the least effective because it mixes with the grease to form a hard soap, which in itself can block the flow of water. Drano is a mixture of sodium hydroxide (which mixes with grease to make a soft soap) and flakes of aluminum. The aluminum reacts to the caustic and forms gas, which loosens the muck. Drano works much better than lye. "99" is all sodium hydroxide. This is what most plumbers use, and this is what I have found to be most effective. It forms a soft soap.

The acid most often used is concentrated sulfuric acid. It is sold under a variety of trade names, including Clobber, Bust-Out and others. It is the most expensive chemical used, costing about four dollars per quart at this writing. It is also the most effective and dangerous. Whereas the alkali types will only work on grease, acid will dissolve grease and hair (including the lint lost by woolen articles).

Precautions—Both the alkali and the acid will "burn" holes in your skin should they make contact and remain. However, the action of both chemicals is insidious. You will feel nothing dur-

133. Always wear rubber gloves when working with acid. Pour it slowly and carefully to avoid any possibility of splashing.

ing the minutes (seconds with acid) that the chemicals require to eat their way through your outer skin layer. When that happens, you are likely to scream. Therefore take extreme care. Wear long-sleeved rubber gloves when working with acid. Keep a nearby faucet open and running while the bottle of acid is open. Do not wait until you feel pain from either the alkali or the acid before you wash the drops of chemical from your skin. If a piece of clothing becomes wet, remove it immediately and wash yourself down. Wash quickly enough and you may escape most of the pain.

Never mix alkali with acid: you will get an explosion. Therefore, if you elect to attack the stoppage with an alkali and it doesn't work—meaning that the alkali remains in the drainpipe—do not add acid. You will have an eruption of hot water, much like an active volcano. The two can be poured into the same drain only if the first chemical has been thoroughly washed away.

Using powder (alkali)—Let the plugged fixture stand unused for as long as possible so that the water remaining is at a minimum. Open a nearby window to let the gas that will be released pass quickly outside.

Dissolve the powder in hot water at a ratio of about ½ cup powder to a pint of water. If all the powder will not dissolve, use more hot water. Never pour the water into the powder, or you will have an explosion. Never pour the powder directly into the stuffed drainpipe. The powder will solidify into one solid, almost insoluble lump. Use very hot water despite the instructions (on some cans) and use lots of caution. The heat speeds the effect of the alkali, but also makes it more dangerous.

Pour the dissolved alkali down the drain. Give it 10 or 20 minutes to do its job. *Do not look into the drain*. The solution frequently boils and shoots back up into the air. Try running the hot water to check for clearance. A fully open drain will accept the wide-open flow of both faucets without any water collecting in the bowl. If the drain is still partially plugged, wait until the fixture is drained and repeat the alkali application.

Using liquid (acid)—Squeeze the drainpipe and the trap connected to the fixture to make certain that the metal is not so thin that the acid can eat right through it. Let the fixture stand as long as you can to permit as much water to drain out as possible. Open a nearby window. Open a nearby faucet and let the water run until you have completely finished with the acid. Put on a pair of long-sleeved gloves. Connect the spout to the plastic bottle of acid as directed.

With your face and head averted, slowly pour about a pint of acid down the drain, taking great care not to splash. Hold your breath until the fumes die down. Do not iook into the drain hole; hot water and acid may squirt up. Remove the spout. Let it remain in the fixture you are working on. Replace the acid bottle's cover. Bring the bottle to the sink or tub in which the water is running. Wash the outside of the bottle of acid. Put the bottle to one side. Return to the plastic spout. Cover its open ends with your rubber-gloved hands and bring the spout to the running water. Wash the spout and your gloves.

Return to the stuffed fixture. Wait 10 or 20 minutes to let the acid do its job. Now try a little hot water. If the drain is completely or satis-

factorily open, fine. If not, repeat the steps. If two one-pint applications of acid do not open the drain completely, you have a solid obstruction inside and will have to open the drain to remove it.

MANUAL CLEARANCE

As an alternative to using the plumber's friend and/or chemicals, to open a stuffed drain you can use a snake to clear it.

Opening a trap—Some traps are equipped with a plug on the bottom of the loop. With the plug removed you can manually reach a portion of the drain.

Start by placing a bucket under the plug to catch the water that will emerge. Use a wrench and back off the plug, taking care not to lose the washer. Fashion a hook from a stiff piece of wire and fish out whatever junk you can. Should the junk be covered with grease you may want to follow the manual cleaning with a chemical cleaning.

Removing a trap—When the trap has no plug the only way you can easily remove the collected debris is to remove the trap from the drain line.

134. When there is a plug at the bottom of a trap, you can easily clean the trap by removing the plug.

135. When there is no plug at the bottom of the trap or if you want to get into the following drainpipe, the trap is removed by backing off the two large compression nuts as shown.

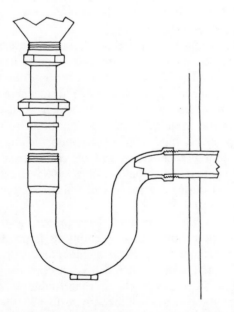

136. When there is only one compression nut, the trap screws onto the drainpipe stub-out. In such cases the sink or basin's tail pipe must be removed and the trap must be unscrewed from the stub-out.

If the trap has two compression nuts, back them both off. Generally, doing so will permit you to pull the trap down and away from the adjoining drainpipes. If you can't, if the drainpipe connecting to the sink or bowl projects into the trap, you have to remove this pipe, generally called a **tail pipe.** Back off the large nut on the underside of the sink or bowl. You will need a spud wrench or large monkey wrench to do this. Then lift the tail pipe and strainer up and clear.

Some traps have only one compression nut. To remove this type, back off the single nut, then raise the tail pipe. Next, turn the trap on its axis. This will unscrew it from the following pipe.

When you have the trap free, it is an easy matter to poke a stiff wire or a snake through it to clear it.

To replace the trap just reverse the steps used to remove it.

Using a snake—Although you can feed a snake down through the drain opening in the bottom of the tub or bowl, it is difficult to make it go through the trap and into the following drainpipe. Therefore, in most instances you are forced to remove the trap or at least its plug, if there is one, in order to get the snake into the following drainpipe.

Removing the trap, unfortunately, doesn't always turn the drain into a garden of Eden. Generally the drainpipe takes a sharp turn a few feet after it has entered the wall. Any debris and grease that has not gathered at the trap will be found at the sharp turn. This is the nature of soil collection; it always accumulates at sharp turns and angle fittings. Guiding the snake past a sharp turn takes patience and effort. Keep poking the snake forward, turning it along its length as you do so. Turning can be more easily accomplished if you keep most of the snake coiled up. Sometimes, bending the tip of the snake helps get it around a turn.

OPENING BRANCH DRAINS AND SEWERS

Each fixture drain is followed by a trap, which leads the water and muck to a branch drainpipe. The branch drain is in turn connected to the soil stack, which is a large-diameter vertical pipe. Its upper end is open to the sky; its lower end makes a turn and crosses the cellar on its way to the house trap. The horizontal portion of the soil pipe is called the house drain. The portion of the pipe that leaves the house trap and continues outside of the house and finally connects to the city sewer main is called the sewer pipe.

In theory, the choice of drainpipe diameters for each section of the drain system is such as to preclude clogging. Practice, however, rarely supports theory. Sooner or later, most if not all drainpipes clog up. The point of blockage varies with individual house design, temperature and the nature of the substances causing the trouble. However, you can assume with fair accuracy that the material accumulates first at the main trap, secondly wherever the drainpipe makes a sharp turn and now and again wherever a drainpipe is exposed to cold air.

Preventing stoppage—Keeping large objects such as toys, silverware and plastic bottle caps out of the drain system is the first and major preventive step. The second consists of giving each fixture drain a shot of alkali or acid now and again. Just how much and how often is difficult to estimate, but it certainly should be applied on a regular basis and well before any slowdown in drainage is evident. Try pouring a water solution containing half a cup of lye down the kitchen sink every month and an equal amount down the other fixture drains every two months. Give the lavatory and bathtub a half cup of acid every six months to get at the hair.

To save yourself the trouble of dissolving the lye you can pour the powder near the drain opening and let the hot water flow over the lye and so carry it down into the drain.

Septic tank bacteria, which do the dirty work of digesting and converting soil to liquid, are adversely affected by strong acids and alkalines. Therefore, if your home drains into a septic tank, dump a few spoonfuls of lye down the various drains every week, rather than a half cup every so often, and forgo the acid treatment. It is too dangerous to open and close a bottle of acid every week. Small quantities of alkali (and acid) are neutralized before they reach the septic tank. Large quantities may not be.

A third preventive step consists of making certain that the drainpipes are not exposed to cold air, particularly the kitchen drain and the branch drain into which it empties. As long as the grease and oils flowing down the drain are warm they remain liquid, or nearly so. As soon as they are chilled they coagulate. If your kitchen drain passes near a window or an exterior wall, wrap it in a glass-wool insulating batt (use wire to hold it in place) to keep it warm.

The final preventive suggestion is not to pour any oil or grease down the kitchen sink. Dump them on your compost heap or garden instead. They sink quickly into the earth and are decomposed into plant food.

Recognizing main-line stoppages—Whenever two or more fixtures are plugged at the same time, chances are good that the branch into which both are connected is blocked. For example, if you use the lavatory and water appears in the nearby bathtub, the drain to which both fixtures are connected is plugged, not the individual fixtures. If using a second-floor fixture results in water appearing in the first-floor bathtub, the blockage is in the line below the first floor.

When water pours out of the fresh-air inlet at the side of your home, the house trap and/or the sewer line behind it are plugged. The fresh-air inlet is the fat pipe that leads outdoors from the house side of the house trap.

Opening a branch drain—If you can, trace the drainpipes leading from the two or more fixtures that are plugged. Very often branch drainpipes are equipped with cleanout plugs where the branch makes a sharp turn, which is usually where the trouble is. Second-floor branch drains are probably hidden in the first-floor ceiling, but first-floor branch drains will be visible from the basement or cellar below—if they haven't been covered during the process of finishing the basement.

If and when you find the cleanout, give the drainpipe as much time to empty as is practical. Then prepare yourself for a minor flood by getting some buckets. When ready, unscrew the plug with a Stillson wrench. If the trouble is right there, you can use a hooked wire to pull the accumulation out. After you do, run your snake up toward the fixtures and then down toward the sewer to make certain that the line is open both ways.

If there is no cleanout plug, or if it cannot be found without tearing up a ceiling or a wall, return to the plugged fixture(s), remove the trap and send the snake down toward the obstruction. As stated previously, you may have trouble in many cases getting the snake past the first sharp turn, but there is no easy alternative, so keep at it.

Opening the house drain—When the first-floor bathtub fills with soil but no water comes out of the fresh-air inlet, your house drain is probably plugged up. The most likely place is where the vertical drainpipe, the soil stack, makes a 90-degree turn and runs in a horizontal direction. The next likely place for debris to gather is wherever the house drain makes a sharp turn. You should find cleanout plugs at both these places. And, if it is more than 45 feet from the bottom of the soil stack to the house trap there should be a cleanout plug somewhere along the middle of the house drain.

Start by shutting off the water the night before and waiting until morning to permit as much water as possible to seep out of the system. Then open the highest plug first. Remove what you can and explore both ways with your snake. Following, go to the next plug down the line and do the same. When you have cleaned out the drainpipe in both directions, replace the plug(s), using plenty of pipe dope on the threads to help seal the joint and make it easier to remove the plugs next time.

Opening the house trap—When soil and water come out of the fresh-air inlet the drainpipe is plugged following the point at which the vent pipe is connected. This is usually the house trap. Again, before opening the trap, give the retained water as much time as possible to seep out and prepare yourself with buckets and mops.

If there are two plugs on the trap, open the one on the sewer side. If no water comes out, the dirt may be at the bottom of the trap or up the pipe a short distance—generally at the lip of the inside of the trap. Take a short piece of wire and carefully poke a small hole through the obstruction. Your purpose is to permit a little water at a time to drain through and so prevent a flood. When all the water has passed, you can hook the end of the wire and remove the blockage.

If there is no blockage at the bottom of the trap, remove the other trap plug. Look for blockage at the entrance to the trap. This is where de-

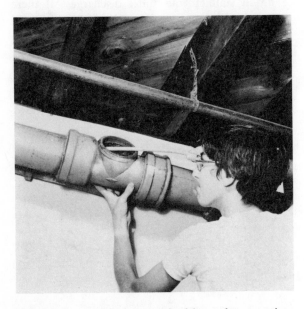

137. Cleanout plugs are required by code every time a drainpipe runs more than 45 feet from the bottom of the soil stack to the house (running) trap. The opening permits you to clear the drainpipe between the stack and the trap.

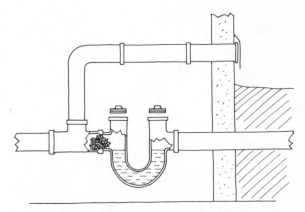

138. Typical house trap equipped with a fresh-air vent (air inlet). When water and muck flows out the vent you know that the trap or the sewer line beyond is plugged up. Muck usually collects at inlet side of house trap.

tergents like to harden and accumulate. Use the same trick of poking a small hole through the accumulation to drain the water off before you remove the obstruction. Clean as thoroughly as you can and use plenty of pipe dope when replacing the screw plugs in the trap.

Opening the sewer—If you have a flood when you open the house-trap plug nearest the sewer line, the sewer is plugged up. Wait until the flood has subsided and see what you can do with your snake. With a husky snake you may be able to open it. If not, you need a thicker, stiffer and possibly longer snake. These can be rented. The best and easiest to use are the power-driven snakes.

Using a power-driven snake—Bear in mind that the electrically powered machine may be grounded, which means that you can electrocute yourself should you be standing in a puddle of water and touch the machine. To preclude this possibility, get rid of all the standing water and put on a pair of heavy, rubber-soled work shoes. With these you can ignore moisture on the cellar floor.

Bear in mind that the motor is geared way down, which means that it develops tremendous torque (turning power). Should you get an arm or leg caught in the machine it will not even pause as it tears the limb off. So, always wear heavy work gloves. Always keep the control switch under your foot so that you can instantly cut the power when necessary. Keep the area clear of all obstructions and junk. If your sewer opening is up near the ceiling and the machine perforce is on the floor, get a helper to hold the machine and work the switch.

Connect a point tip to the snake end. Hand push the snake into the sewer as far as you can. Turn the machine on and carefully hand push the snake forward (the better machines do this for you). Should the snake get stuck, shut the machine off and reverse its rotation. When you have driven the snake all the way through—and some sewer lines run 100 feet before they hit the city main—retract the agile beast and put the cutting tool on its end. This will cut roots and the like from inside the sewer pipe as you guide it through.

Withdraw the snake and have some helpers flush all the toilets at one time. Use a flashlight and peer down the trap hole to see how well the sewer takes the flow. There shouldn't be any backup. A 4-inch drainpipe will easily take half a dozen simultaneous toilet discharges. If there is a water backup, run the snake and cutting tool back down the sewer pipe.

CHAPTER TWELVE

Replacing Faucets and Fixtures

When you replace a faucet or a fixture with another, identical unit, the work is minimal. When the replacement is different from the original, whether it is more modern or even older, problems can arise that will make the work much more difficult than can be readily imagined. To forestall this possibility, always plan the positioning and connection of the faucet or fixture carefully even before you make any purchases. This can be done with the aid of manufacturers' literature or by measuring the faucets and fixtures while they are still in the shop.

Measure the old and the new and make certain that the new sink, let us say, will fit into its allotted space; that its drainpipe can be connected without ripping down a wall or rebuilding an entire run of drain piping. Make certain that the new faucet will fit the hole or holes in the old sink before you start taking things apart.

REPLACING INDIVIDUAL FAUCETS

For the purpose of removal, we can divide individual faucets into two groups: screw-on and fitting-connected. One type can be differentiated from the other by peeking behind the fixture. If there is nothing but a straight pipe visible, it is a **screw-on faucet.** If you see a large and a small nut on the faucet shank, it is a **fitting-connected faucet.** Should the fixture be close to the wall and impossible to see behind, try pulling outward on the faucet. If it moves even a little, chances are that it is a screw-on. If it doesn't budge it is probably fitting-connected.

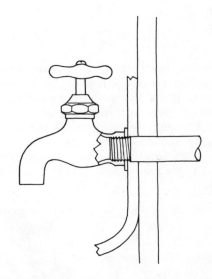

139. Old-time, screw-on faucet. It is simply unscrewed. Before attempting to remove it, be certain that the sink isn't being held up by the pipes alone.

Screw-on faucets are removed by simply unscrewing them with the aid of a wrench. The replacement faucet must have an identical, female-threaded end. Some pipe dope is smeared on the male-threaded feed pipe, then the new faucet is screwed on. In doing so, take care to quit turning when the faucet is fairly tight against the fixture and in its proper vertical position.

Fitting-connected faucets cannot be unscrewed. They usually have nibs on their shanks that prevent them from being turned. If you force the

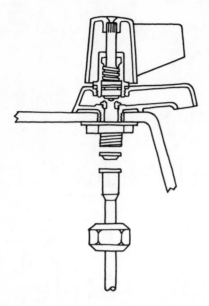

140. Cross-sectional view of a fitting-connected faucet. This faucet cannot be removed by unscrewing it from its supply or feed pipe. To remove faucets of this kind you must get behind them and unscrew the large nut holding the faucet to the fixture. If you are dealing with a dual faucet—two faucets integral with a single casting—you must remove the feed pipes and nuts holding both faucets to remove the assembly.

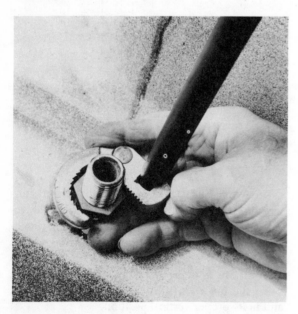

141. Working end of a basin wrench. Hook-head is on a swivel and can be used for removing and tightening shank nuts otherwise unreachable.

faucet to turn you may crack the supporting fixture.

To remove a fitting-connected faucet you must get behind the fixture; unscrew the compression nut holding the feed pipe to the faucet's end, and then unscrew the larger nut holding the faucet in place. After the two nuts are backed off and removed the faucet can be lifted out of place.

If your work space is limited you will need a basin wrench to remove the nuts. If the large nut is corroded fast, try a hacksaw or a hacksaw blade alone. Do not try to chisel the nut free, as porcelain and porcelainized steel are easily chipped and cracked.

Fitting-connected faucets are replaced by simply inserting the new faucet into the old hole and connecting the feed pipe after the holding nut has been tightened down. If the individual faucet mounts on wood (even wood covered with plastic) coat the faucet shank and flange with a thin layer of soft plumber's putty to seal out water.

Feed-pipe problems—If the feed-pipe termination matches the end of the replacement faucet, no problem. If it doesn't, the feed-pipe termination has to be changed. This is discussed a few pages on, under the heading Water-feed Connections.

REMOVING DUAL FAUCETS

Dual faucets are always fitting-connected, so there is no way to remove these faucets except by getting behind them and removing the two nuts on each shank.

Surface-mounted duals—You will find this type of faucet atop a kitchen sink or a bathroom

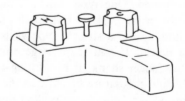

142. Typical surface-mounted dual faucet. Remember, you must close both the hot and cold feed pipes when removing either faucet stem for repairs.

lavatory. The body of the faucet is always clearly visible atop the fixture. These faucets can be removed by simply loosening and removing the pipe and the nuts on the underside of the fixture, then lifting the dual faucet up. Where there is a sink spray-hose, the hose can usually be readily disconnected from the faucet by simply unscrewing it where it passes up through a central hole in the fixture.

Under-surface duals—You will find this type mainly in lavatories. You will recognize them by the visual absence of the faucet body. Only the faucet spout and handles appear above the top surface of the fixture.

To remove an under-surface dual faucet the handles are removed. Next, the large nuts that are now visible are removed. In some designs you will have to remove the packing nut and stem to get at the large nut that holds the faucet in place. Next, protective tape is wrapped around the spout and it is unscrewed with the aid of a wrench. In some models the pop-up valve control behind the spout will slip right out with the spout removed. In others, the valve control mechanism has to be disassembled. Next, the feed pipes are

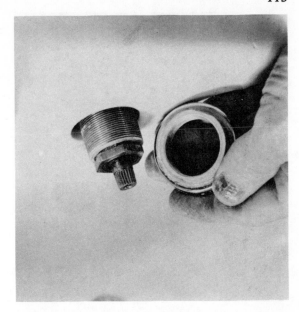

144. Large nut has been removed.

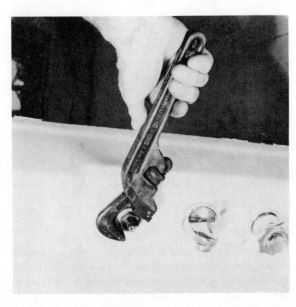

143. Removing an under-surface dual faucet. Handle has been removed; now large nut (round on this model) is being removed.

145. Spout is unscrewed after being wrapped with tape to prevent marring surface. Pop-up valve control will slip off when spout is removed. When faucet feed pipes are disconnected, faucet can be moved backward, down and free of the sink or the lavatory.

146. Ground-joint fitting used to connect water feed to faucet shank. Note that fitting is a short length of tubing that is screw-coupled to threaded feed pipe. Faucet shanks designed to accept ground fittings will accept no others (they leak when used with others).

the shanks and connect the feed pipes. (Feed-pipe connections are discussed a few paragraphs on.)

Under-surface duals—The replacement faucet must be an exact or a nearly exact replacement because there is generally very little room to spare beneath these fixtures.

Start by slipping the faucet body casting into place. Then assemble the pop-up valve mechanism if there is one. Next, connect the feed pipes; then replace the shank nuts and the handles. The reason for this sequence of reassembly steps is that it is usually easier this way.

WATER-FEED CONNECTIONS

There is no problem with water-feed pipe connections when you install a screw-on faucet. It is

disconnected. In some models the faucet body casting can now be moved back, down and out. In others, the pop-up valve levers have to be taken apart to free the faucet casting.

REPLACING DUAL FAUCETS

Surface-mounted duals—The shank spacing on the replacement unit, be it older or newer, must match the holes in the top of the sink, counter or lavatory. However, you do not have to replace a spray faucet with another spray faucet if you do not wish to do so. Simply ignore the central hole. The spray hose can be removed or simply left in place, unconnected.

Start by cleaning the surface of the fixture. Next, cover the bottom of the faucet with a thin layer of soft plumber's putty or any soft putty. This is necessary to seal the bottom of the faucet to the fixture or counter top and to prevent water from entering. Next, run the holding nuts up on

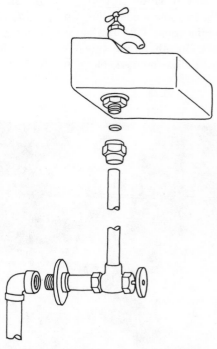

147. Slip fittings are found on old plumbing installations, most often connecting the feed pipes to a toilet tank. On occasion you will also find them used with old sinks. If the faucet or ball cock valve is replaced, you have to replace the slip fitting with a Speedee. Slip fittings will not fit new faucets and valves.

screwed directly onto the pipe that feeds it. Fitting-connected faucets are another matter, as there are several different fittings that have been used to connect the water supply pipe to the end of the faucet. These fittings are not interchangeable. If your new faucet does not accept the existing feed-pipe terminations or fittings, you have to change the old fittings to make the connections. The most frequently encountered feed-pipe-to-faucet-end fittings are described following.

Slip fittings—These consist of a feed-pipe end that fits loosely inside the end of the faucet, a slip washer that goes between the inner pipe and the faucet end and a compression nut. The nut screws onto the shank of the faucet and holds the slip washer in place, sealing the joint. The distance the feed pipe enters the shank of the faucet can be varied somewhat.

Ground-joint fitting—A short section of pipe with one end ground to a smooth cone fits into a similarly shaped opening in the end of the faucet shank. A compression nut behind the ground joint holds it firmly to the faucet end and seals the joint. The feed pipe must be in nearly perfect linear alignment with the faucet shank and the ground joint must be close enough to the shank end to be pulled into place by the compression nut.

Metal-ring, compression-joint fitting—This fitting can only be used with copper tubing feed pipes. The fitting consists of a compression nut which is slipped up onto the end of the feed tube (see illustration 12). Next a metal ring is slipped over the tube end and both tube end and ring enter the opening in the end of the faucet. Bringing the compression nut up and tightening it against the metal ring by running it up on the threads compresses the ring into the space between the tube and the faucet-shank opening. The trouble with this arrangement is that the compression nut tends to distort the metal ring. The joint sometimes leaks when merely taken apart and remade with the original faucet. In such cases there is nothing to do but install another length of copper tubing and a new ring.

Speedee fitting—This is a length of ⅜-inch copper tubing with a preformed, ball-shaped end. The tubing may be ridged for easier bending (better but costlier). The ball fits into the end of

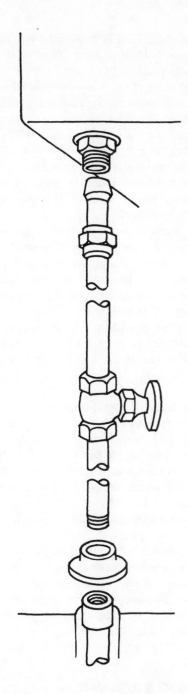

148. A ground-joint fitting used with a ball cock valve. If the joint between the fitting and the bottom of the ball cock valve or faucet leaks, try wrapping a few turns of waxed string around the feed pipe behind the ground joint, then tightening the locknut. If that doesn't do it, try drying the joint and sealing it with self-forming automobile gasket cement.

the faucet shank and is followed by a compression nut that holds it in place. The Speedee is the latest in the plumbing industry's efforts to facilitate faucet connecting. It is used with all new faucets that are not directly soldered to the water supply pipe. The Speedee can be removed and reconnected countless times without difficulty.

Changing the water-feed connection—The task of changing the water-feed connection to match the needs of the new faucet is relatively simple if you have the space to work in. If the space is limited, it can be difficult.

To go from a slip fitting to a ground-joint fitting the old pipe must be removed, cut and threaded to accept the ground-joint fitting. If the lengths are right you can sometimes effect the change by using couplings and short nipples without cutting and threading the old pipe. The criterion is that the ground-joint fitting must terminate within a fraction of an inch of the faucet end and the pipe that carries the fitting must line up with the faucet's shank. This requires careful measurement.

To go from a metal-ring compression-joint fitting to a Speedee fitting the old copper tube is cut short and joined to the Speedee with a sweated-on coupling. When you do this, bend an S shape into the Speedee tubing. This will permit you to easily adjust the Speedee end to fit the faucet connection. Don't forget that the compression nut must be on the connecting tubing before you make the solder joint.

To go from a slip fitting or a ground-joint fitting to a Speedee, the old pipe must be cut and threaded, or replaced by a short nipple and coupling. The Speedee tube end is sweated to an adapter fitting, the other end of which screws onto the pipe end.

REPLACING A SPEEDEE

Sometimes an existing Speedee is damaged when the fixture to which it is connected is replaced. Sometimes an existing Speedee is too short to reach the faucets on a new fixture. In such cases the Speedee must be replaced.

When the non-faucet end of the Speedee is soldered to the water supply pipe, the Speedee is

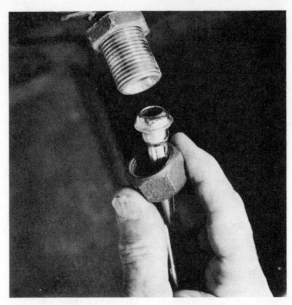

149. Speedee fitting, the fitting most often used on modern faucets, being assembled.

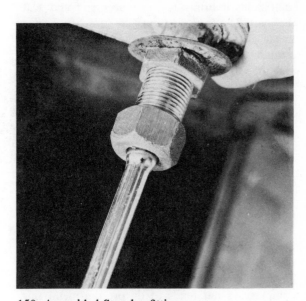

150. Assembled Speedee fitting.

merely unsoldered and the new one is soldered in its place. When the Speedee's end is connected to a shutoff valve by a metal-ring compression joint, removal is easy but replacement requires a little care.

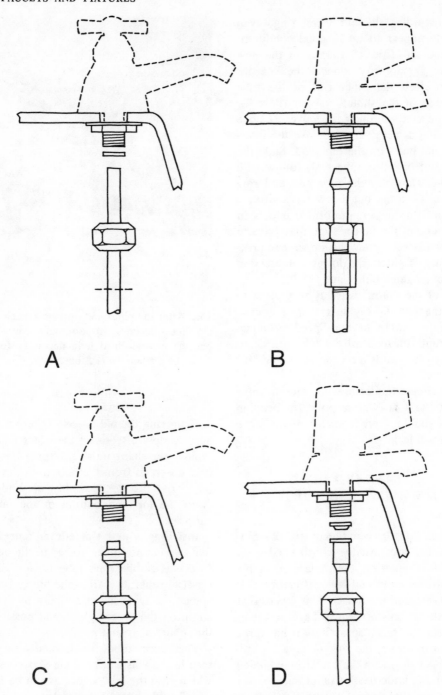

151. To change a slip-fitting connection (A) to a ground-joint connection (B), cut the feed pipe and thread it. Then connect a ground-joint fitting to the original feed pipe with a threaded coupling.

To change a metal compression-ring connection (C) to a Speedee connection (D), cut the copper-tubing feed pipe. Then solder a Speedee fitting to the end of the copper tube. If you put an S bend in the Speedee tube you will find it easier to connect.

Start by discarding the old metal ring, even though it may appear to be in good condition. Carefully bend an offset (S shape) in the new Speedee tube. Temporarily connect the Speedee to the faucet. Rest the other end of the tube against the side of the shutoff valve. Allow for the ½ inch the tube will enter the valve. Remove and cut the tube as necessary. Remove the internal and external burrs on the tube end. Slide the compression nut onto the tube end; follow with the metal ring. Now slip the tube end and ring into the valve. Give the nut a few turns. Rotate the tubing until the upper, preformed end is in line with the end of the faucet (you may have to "spring" it in place). Screw the upper compression nut home. Tighten the lower compression nut. Do not overtighten either nut.

If the top of the shutoff valve is very close to the end of the faucet—say less than 1 foot—carefully bend a circle in the Speedee tubing. This will permit you to bend the tube a little to easily make the tube ends meet the valve and faucet connections.

Under no circumstances use the shutoff valve or the faucet end as a fulcrum point for bending the tubing. If you do, there is a very good chance that the joint will leak.

REPLACING A STANDARD DUAL FAUCET WITH A SINGLE-LEVER FAUCET

Make certain that the new faucet will fit—Single-lever faucets are usually constructed with both feed tubes entering the faucet assembly directly below the spout and lever assembly, that is to say, the center of the faucet. This means that your old sink or lavatory must have a center hole. If the old unit has a spray hose it has three holes. If it has no hose, the sink or lavatory still may have three holes, with the center hole hidden and unused. Look underneath to find out.

Depending on the design of the single-lever unit you purchase, the spread of the two outside holes may or may not be important. Usually, the single-lever faucet has two movable mounting bolts which can be adjusted to the usual run of hole spacings; but check first. Again, you can do this from the underside.

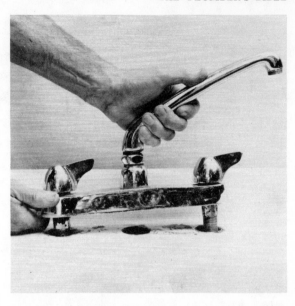

152. After the pipes and nuts have been removed the old faucet is simply lifted up and removed. This sink has the central third hole necessary for the installation of a single-lever faucet.

Removing the old faucet—There is no problem here. Simply disconnect the water feed pipes and remove the shank nuts. The spray hose, if there is one, unscrews from the bottom of the faucet casting. Now lift the assembly up and out. Then clean the top and bottom of the faucet-contact surfaces.

Installing a new single-lever faucet—You will find two copper tubes brazed to the bottom of the faucet assembly. One tube is carefully given an offset to separate it from the other. Do not simply spread the two tube ends apart, as that will put a strain on the brazed joints and possibly will kink the tubing near that point.

The faucet tubes are normally sweated to the feed lines. This can be done above the fixture, in which case the feed lines have to be long enough and flexible enough to be brought a few inches above the center hole in the fixture. Or, if there is sufficient work space, the faucet tubes can be sweated to the supply lines beneath the fixture. The first method is much easier, but you have to be careful not to kink the tubing when you bend it down and out of the way.

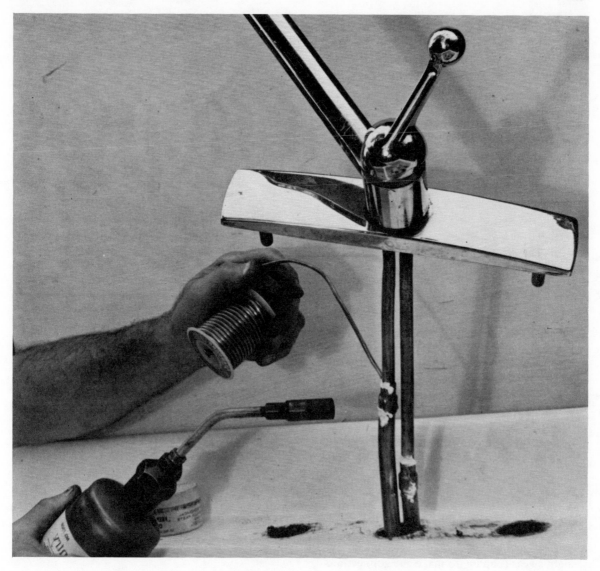

153. If you make the copper feed tubes a little longer than necessary, you can do your soldering above the sink, rather than underneath it.

If the feed is pipe, remove it and cut it shorter. Thread the cut end and attach an adapter fitting that will take the faucet tubing. Or discard the old pipe and replace it with a nipple and go from the nipple to the faucet tubing with an adapter, a length of tubing and a sweat coupling. Use ½-inch tubing if the distance from the nipple to the faucet tubing is more than 2 feet. Use ⅜ if you wish, if the distance is less than 2 feet. Use ½-inch tubing for any tube length if you are connecting a sink or tub. Smaller-diameter tubing will reduce the flow of water too much.

154. Cover the bottom of the new faucet with a layer of plumber's putty before you seat it. Then, when the bolts have been made tight, clean off the excess putty with a screwdriver or a knife.

Should the original water feed be ½-inch tubing, clean up the ends and use a reducing sweat coupling to connect the feed tube to the faucet tubing. If the tubing is ⅜-inch you can use a straight sweat coupling to make the connection.

If the original feed is a Speedee fitting, which is ⅜-inch tubing, you have the choice of cutting the formed end off and using a sweat coupling or disconnecting the Speedee and working from there. Generally, it is easier to remove the Speedee, because it is made from rigid tubing and is not easily bent without kinking.

Cover the bottom of the new faucet with a layer of plumber's putty or soft putty. Now sweat the feed tubes onto the faucet tubes. This done, press the faucet down into place, taking care not to kink any of the tubing, and run the nuts up on the bolts on the underside of the faucet. Remove excess putty with a knife and the job is done.

Should you get hot water with the lever in the cold-water position, you have reversed the feed lines.

REPLACING A WALL-HUNG SINK OR LAVATORY

Removing the old fixture—Start by disconnecting the drainpipe and removing as many sections as you can to facilitate fixture removal. Then disconnect the hot- and cold-water feed pipes. If you are working on an old-time fixture and the faucets are screw-ons, disconnecting the faucets means simply unscrewing them. However, as the wood carrying the fixture-supporting bracket may be long since rotted away, do not unscrew the faucets until you have two helpers standing by: the fixture may be hanging from the pipes instead

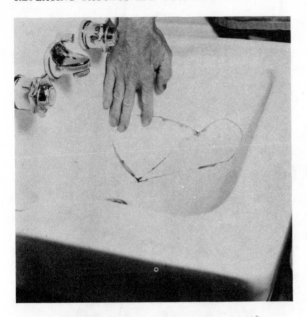

155. This lavatory has seen better days. The cracked portion is held in place with epoxy cement, but the surface is rough.

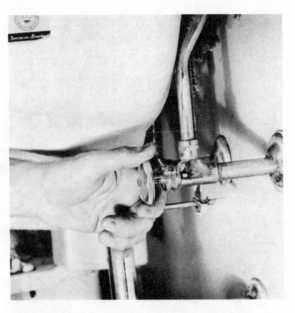

156. The first step in removing and replacing a lavatory is to close the water supply shutoff valve.

157. Disassemble the pop-up valve mechanism.

158. Disconnect the water feed pipes. This is a metal-ring compression fitting (joint).

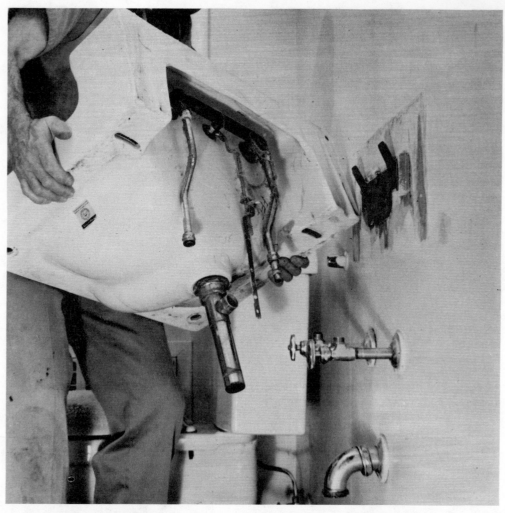

159. The tail pipe is disconnected from the trap. Now the entire assembly is lifted. As can be seen, it was unnecessary to disconnect the pop-up valve, as it was easier to separate the tail pipe from the trap than the tail pipe from the lavatory.

of from its bracket, in which case it will likely fall to the ground unless you manually support it.

With the feed pipes and the drain disconnected, the fixture can be lifted up and away from the wall. Whether or not the old fixture is hanging from its pipes, it is wise to have a helping hand when removing a wall-hung fixture. Many are much heavier than they appear to be.

Check the existing fixture bracket—Examine the existing fixture bracket or hanger, as it is sometimes called. Make certain that it is secure

and sturdy. If you have help and have the new fixture handy, try hanging the new fixture on the old bracket. See if the fixture is flush against the wall, and if the bracket rests deeply in the pockets made for it. Check the fixture's height from the floor. Its surface should be about 31 inches above the finished floor. If all is well you can go right ahead and connect it up.

On the other hand, if the old wall is partially rotted, if the new fixture is of an entirely different size and shape and you do not know whether or

160. The screws holding the fixture bracket are tightened. If the replacement fixture is not identical to the old unit, you must make certain that it fits on the bracket properly.

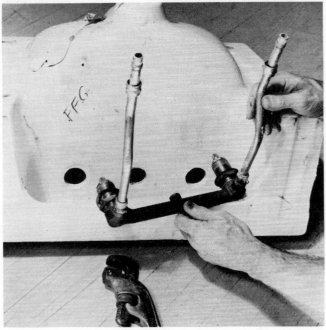

161. The faucet assembly is disconnected and removed. This is most easily done with the lavatory upside down and resting on the floor.

not it will fit the space or if you want to move the new fixture a distance to the right or left, the following procedure is recommended.

Check the bracket—With the new fixture resting on the floor, hold the bracket, new or old, against the rear fixture edge. Check to make certain that the bracket fits snugly and properly.

Measure the new fixture—Measure the major dimensions of the new fixture and then measure the allotted space to make certain that it will fit with sufficient clearance. Next position the bracket against the fixture and see how many inches the screws in the bracket are above or below the surface of the fixture (the front ledge top is considered the surface or top of a fixture).

Mount the bracket—Use the measurement you just made and draw a long horizontal line on the fixture-supporting wall. Eventually the screws holding the fixture bracket will pass through this line. Therefore, set the height of this line so as to bring the surface of the hung fixture the usual 31 inches above the finished floor. For example, if the screw line on the bracket is 2 inches above the surface of the fixture, then the line must be 33 inches above the finished floor.

Hold the bracket up against the wall with its screw holes along the line and the bracket the correct distance from the nearest side wall. Mark the screw holes. Drive test nails into the wall. If you do not reach solid studs (vertical 2×4's), you must rabbet a cross brace into the studs to support the fixture. Cut a hole in the wall long enough to more than reach the two nearest studs and roughly centered horizontally in relation to the desired fixture position. Make the hole about 5 inches wide. Remove the plaster or Sheetrock. Cut two notches in the studs to accept a 1×2 or a 2×4 cross brace. Nail the brace firmly in place; then wet down the edges of the hole and fill it flush to the wall surface with a strip of Sheetrock and plaster.

Next the bracket is fastened to the cross brace with wood screws long enough to go well into the wood. The bracket is now correctly positioned horizontally and vertically. You can hang your new fixture in place now.

Connecting the drain—Keep these points in mind when connecting your newly hung fixture to the existing drainpipe.

You must include a trap. If you don't, an odor will constantly emanate from the fixture's drain.

The **stub-out** (where the drain enters the wall) must be lower than the bottom of the fixture's sink or basin by a few inches. If it is not (because the new basin is too low or the old was too high) the new fixture will not drain properly.

The drainpipe leaving the bowl should be the same size or smaller in diameter than the drainpipe at the stub-out. If it is not there will be a tendency for debris to pile up at the stub-out because of the reduction in flow rate caused by the reduction in pipe diameter.

The connection between the fixture basin and the stub-out should be as simple and as direct as practical. Every additional turn slows the flow of soil and increases the frequency of drainpipe plugging.

Use a trap with a cleanout plug on the bottom. It is a little more expensive, but it can save you a lot of fuss when the drain clogs, as all drains eventually do.

The actual connection between the fixture drain and the stub-out can be made in a number of ways. In some instances the end of the fixture trap or the fixture trap extension is simply screwed into the female end of the stub-out. In other instances the stub-out-to-trap connection is made with a flexible-ring compression joint. When the stub-out is copper tubing the joint may be sweated. When it is plastic, the joint can be solvent welded. The method used is unimportant so long as a good joint is made.

Making the drain-to-fixture connection—To swing to the right or left from the stub-out you can use a trap extension having an elbow and joined to the trap by a compression joint.

To adjust the fixture's elevation in relation to the trap you can use a compression joint on the fixture's tail pipe (the pipe leading downward from the drain hole in the basin). The compression joint and a pipe of the correct diameter will enable you to slip the tail pipe upward and downward a few inches as might be necessary.

The fixture's tail pipe is part of or attached to the drain basket, which is what you see when you look down the drain hole in the fixture. The basket is mounted either with the provided gaskets above and below the metal-to-basin contact area,

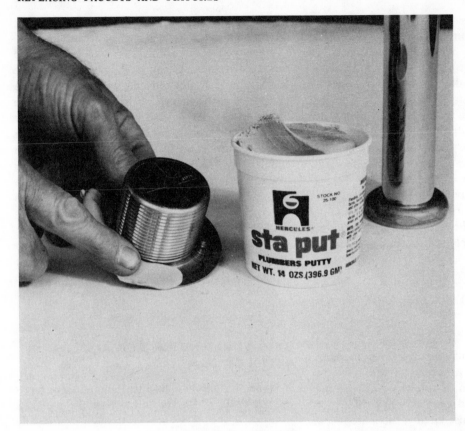

162. Installing a drain basket and tail pipe on a sink. If you have no washer or gasket, a layer of plumber's putty is used instead to insulate the flange from the porcelain.

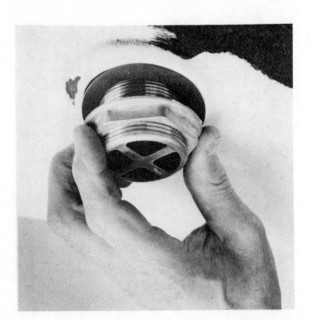

163. A gasket or putty, but preferably a gasket, is used on the underside of the sink to insulate it from the large nut.

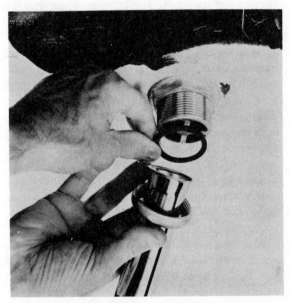

164. A second gasket is installed *between* the lip of the tail pipe and the bottom of the drain basket. The tail pipe is next attached.

or using two thin layers of plumber's putty to prevent the metal from making direct contact with the porcelain or plastic. You will need a large Stillson wrench or a spud wrench to take up on the basket nut. Just don't make it too tight.

Installing the faucets—The faucets, whether single or dual, can be installed now, with the fixture hung and the drain connected. Or, they can be installed before the fixture is hung. Depending on the nature of the fixture, one approach is sometimes easier than the other. Sometimes it is easier to attach the feed pipes to the faucets before the fixture is hung, and then connect the lower end of the feed pipes to the shutoff valves of the water supply stub-outs.

REPLACING CABINET SINKS AND LAVATORIES

There are two kinds of cabinet-type sinks and lavatories. One kind may be described as **unitized.** The bowl or basin is integral with the top of the cabinet. You can, if you wish, lift the entire cabinet top off and carry it away. Many of the older apartment house kitchen sinks are made this way. The other kind may be described as **counter-top mounted.** The bowl or basin and the faucet(s) fit into holes cut into the counter top and can be separated from the counter top.

Both types of cabinet sinks and lavatories are handled exactly like wall-hung fixtures. Everything that was stated regarding wall-hung fixtures applies to self-supporting cabinet-type fixtures. There is only one minor difference from a plumbing point of view. Cabinet-type fixtures are easier to replace. You don't have to worry about the wall bracket. The cabinet hides unsightly pipe work, and if you do not fasten the cabinet down until you have connected all of the pipes, you can move the cabinet about a bit to accommodate your piping.

Mounting considerations—Unitized fixtures are prefabricated. The bowl is part of the cabinet top. Faucet holes are predrilled, so you can't go wrong.

On the other hand, although counter-top fixtures are sometimes sold with the top already cut and the parts all mounted, sometimes you have to cut your own holes (for example, if you are fitting a sink to a new kitchen cabinet).

When you plan to cut your own holes, make certain that there is sufficient counter-top space to accommodate not only the basin, but also the underside of the faucets. You need room for the fittings. And note that the holes must be cut very accurately, especially for a basin that fits within a ring that, in turn, fits within the hole in the counter top. The inside and outside of this ring or oval must be given a thin coating of plumber's putty before it is installed. This is done to keep water from getting inside the cabinet and wetting the wood.

The basin types that rest atop the counter have a lip that overhangs; they do not require the hole be as accurately cut. But you still need to seal the joint between the lip and the counter top with putty.

REMOVING AND/OR REPLACING A TOILET

Drain the toilet—Close the shutoff valve feeding the tank. Flush the toilet. Remove the tank

165. The first step in removing a toilet consists of emptying it of water. Flush the toilet, pump the water out of the bowl with a force pump and drain the last of the water out of the tank with a sponge.

cover and put it carefully aside—it is easily broken. Take a sponge and remove all the water remaining inside the toilet tank. Take a plunger and drive as much water as you can down the drainpipe. Whatever water remains in the toilet's trap will later spill out onto the floor, so keep a rag handy.

Remove the tank—First disconnect and remove the water feed pipe. If the toilet is an old one with a brass elbow connecting the tank to the bowl section, remove the elbow. Then, with someone or something holding the tank up, remove the two brass screws holding the tank to the wall.

If the tank rests atop the bowl section, loosen and remove the two brass bolts holding the tank to the bowl. You will see their heads inside the tank, at the bottom.

Remove the bowl section—Look at the bottom of the bowl section. You will see two or four porcelain caps protruding from the base. Wrap some tape around each cap and use a Stillson wrench to remove them. Don't be surprised if they crack, which they probably will if plaster instead of

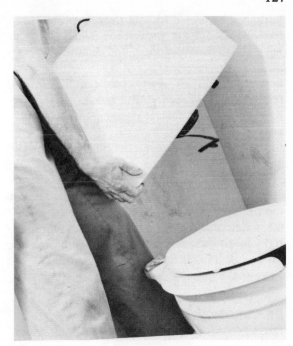

167. Now you can lift the tank up and away.

166. Disconnect the tank from the water supply. Shown is a close-up of the special Speedee connection most often used between the tank water supply and the bottom of the ball cock valve. Loosen and remove the bolts holding the tank to the bowl.

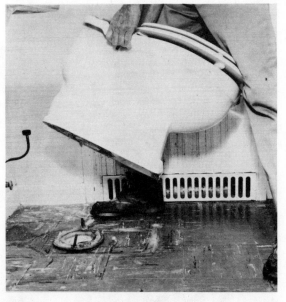

168. Loosen and remove the hold-down bolts on either side of the bowl. This done, you can lift the bowl off and remove it.

putty has been used to hold them in place (don't fret—new ones are available). When the caps have been removed, loosen and remove the brass nuts that are revealed.

Now, if you are ready, you can lift up and remove the bowl section. But before you attempt it, bear in mind that it is more awkward than heavy, and that you have to lift it straight up for about 1 inch before you can move it sideways.

The floor flange and connected drainpipe are revealed when you remove the toilet bowl. The top of the drainpipe is open. If you have "pulled" the toilet to remove an obstruction, simply go right ahead and remove whatever is within reach. To be certain that there is nothing farther down, run a snake through the drainpipe.

REPLACING THE SAME OR AN IDENTICAL TOILET

Examine the floor flange. If there is a gasket resting atop it, examine the gasket. If it is aged, torn or ripped, replace it. Then examine the un-

derside of the toilet. If any bits of the gasket remain adhered to the porcelain, scrape them off.

If there is no gasket on the floor flange, look at the underside of the toilet bowl. You will see a ring of wax there. To be certain that you will make a perfect seal when you replace the bowl, remove the old wax ring and install a new one.

The wax ring fits around the "horn," which is the very short tube that projects downward from the base of the bowl. The horn enters the floor flange and so leads the soil into the drainpipe. The gasket or the wax ring seals the horn to the floor flange—a most important task.

With either the gasket in place on the floor flange or the wax ring on the horn, the toilet bowl is lifted up and positioned directly over the floor flange. The toilet horn must be in line with the opening in the floor flange. The hold-down bolts projecting upward from the floor flange must be in line with the holes in the base of the bowl. When everything is lined up perfectly, the bowl is lowered into place. This is not easy to do, so be prepared for several tries.

Make the hold-down nuts snug. Use putty to

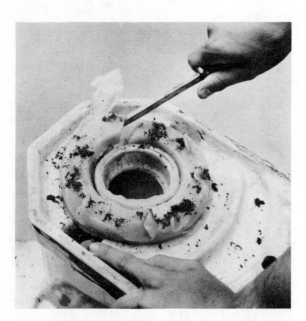

169. Turn the bowl upside down. Remove the remains of the old wax ring.

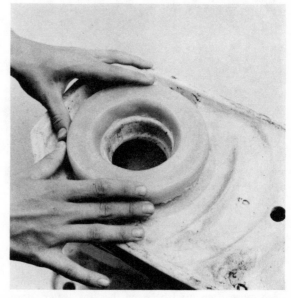

170. Press the new wax ring firmly in place around the toilet's horn.

171. Remove almost all of the old wax left on the toilet flange on the floor. Leave just enough wax to hold the bolts upright. Carefully lower the bowl over the hole, making certain that the bolts are in their proper position before you let go of the bowl.

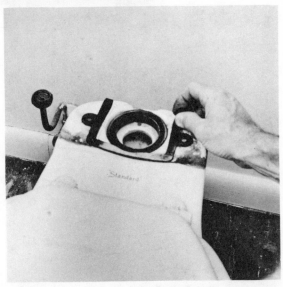

172. Replace the rubber grommets on the bolts that hold the tank in place. On the model shown, a gasket is used instead. The large central rubber washer, called a spud washer, goes back with the thicker side up.

replace the covering plastic caps. Install the large washer that goes between the bowl and the tank. It is called a **spud washer** and the larger side goes up. Inspect the tank bolts. If their rubber grommets are torn, if the bolts are rusted, replace them with brass nuts and bolts. Again, remember that these bolts do not need to be more than snug. It is easy to crack the porcelain tank by overtightening.

INSTALLING A DIFFERENT TOILET

When you replace the original toilet with an identical fixture there is no problem. When your replacement toilet is different from the original you can run into difficulty. The discharge horn on the toilet *must* fit into the floor flange and the drainpipe. This flange cannot be moved without ripping up the floor and reconnecting the toilet drainpipe. Possible, but hardly practical.

To make certain that your replacement toilet will fit the existing space you need to know the exact distance from the center of the floor flange to the rear wall, so that you can compare this to the distance from the center of the horn to a point directly beneath the tank cover on the replacement toilet. Obviously, if the horn-to-tank-

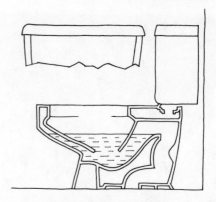

174. A typical manufacturer's sketch. This model requires 12 inches from the center of the toilet flange, plus 2 inches or more of clearance to the rear wall. *Courtesy of American Standard.*

173. To make certain that a new toilet will fit, measure the distance from the center of the toilet flange to the rear wall. Compare the distance with the new toilet or with the manufacturer's literature.

cover dimension is larger than the floor-flange-to-wall dimension the new fixture isn't going to fit without rebuilding the wall.

Measuring clearance—The certain way consists of removing the old toilet and exposing the floor flange. Then you can easily and accurately measure from its center to the rear wall. The easy but risky way consists of measuring to the wall from one of the hold-down bolts on the in-place, old toilet. Risky, because you cannot really be certain what the bottom of the old toilet looks like unless you remove it. However, you may want to measure horn-to-wall clearance this way, so here is how it is done.

If there is only a single pair of hold-down bolts on the bottom of the toilet, measure from one bolt to the rear wall. Usually these bolts are in line with the center of the floor flange. If there are two pairs of bolts, flush the toilet and watch the flow of water to determine if the drainpipe is between the front or rear pair of bolts. Then measure to the rear wall from one of the bolts that you believe is in line with the center of the floor flange.

Connecting the water supply—When the replacement toilet and tank are identical there is no problem; the feed pipe is simply reconnected to the new tank. When the replacement is different you will greatly simplify the job by "locating" the new tank and its ball cock valve connection on

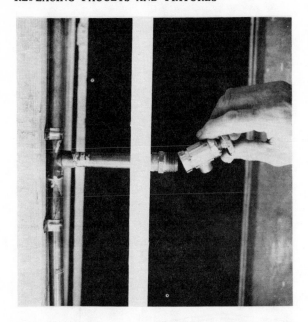

175. To bring the water supply out from a run of copper tubing, use a Tee, plus a short length of tube connected to an adapter. The angle shutoff valve fits the adapter.

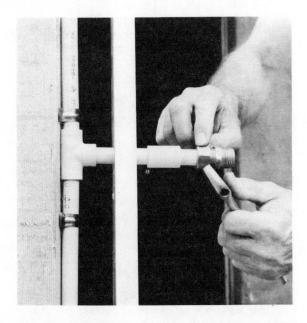

176. To bring the water supply out from a plastic pipe, use a Tee, a section of pipe, a coupling and an adapter, as shown. The angle valve, which connects to the bottom of the Speedee fitting, goes on the end of the adapter.

the wall behind the toilet. When you mark the approximate position of the ball cock valve connection on the rear wall you will know whether or not you can make the connection between the shutoff valve (or stub-out) and the toilet tank. If you can't, because the valve will be behind the tank or too close to it, move the shutoff valve before you install the toilet.

Open the wall, disconnect the shutoff valve and extend or shorten the pipe as necessary. Generally the shutoff valve is positioned beneath the tank, about 1 to 1½ feet down. Whether you are working with copper, plastic or galvanized, be certain to strap the stub-out firmly to a stud or a brace.

REPLACING BATHTUBS

In order to remove and replace a bathtub with a minimum of difficulty, a knowledge of conventional tub-installation practices is necessary. Therefore, how tubs are installed is reviewed briefly prior to discussion of the steps necessary to remove an existing tub.

Tub types and their installation—There are two types of tubs in common use: **free-standing** and **wall-contact**. A free-standing tub has floor space all around. All of its four sides are finished and none touch a wall. A wall-contact tub touches one or more walls. Usually it fits into an alcove; in such cases three of the tub's sides make contact with walls and the fourth is finished.

All tubs normally rest on the subfloor. Floor tiles, asphalt or ceramic, make contact with the sides of the tub. In theory, if you were to disconnect all the pipes from a free-standing tub you could, if you were strong enough, lift the tub directly upward and walk away with it. In practice, this can only be done with an asphalt tile floor. When the floor is ceramic, some of the tile will adhere to the tub sides and the tile floor will be damaged to some extent. The only exception would be a really old tub standing on legs.

Point one: Do not plan on removing a free-standing tub without causing some floor-tile damage.

Point two: Do not plan on removing a wall-contact tub without some wall and floor-tile damage.

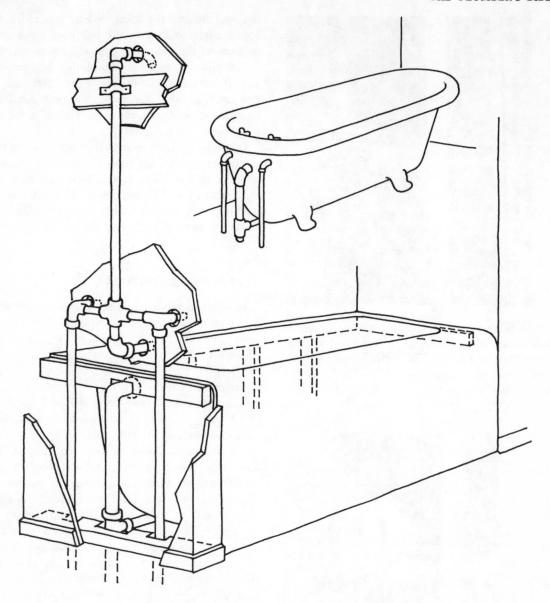

177. Typical tub layouts: old above, new below.

Wall-contact tubs have lips about ½-inch high that rest against the studs (wall frame) and beneath the Sheetrock and wall tile. It is just about impossible to remove a wall-contact tub without removing the row of tile directly above the tub's edge(s), *plus* a strip of Sheetrock along the same edge(s). In addition, you should not hope to remove a wall-contact tub without some floor-tile damage.

MEASURE BEFORE YOU CHOP

If you are replacing a free-standing tub that has legs, all you need worry about are the pipes. The new tub can be placed directly atop the old tile floor. Use some caulking cement beneath the tub's lower edges to seal the joint.

If you are replacing a free-standing tub that

rests on the subflooring and has floor tile butted against its side, you can usually avoid floor-tile repairs by installing a larger tub, which can rest atop the undamaged tile. (Use plaster or cement to level the floor before you install the new, larger tub.)

If you are replacing a wall-contact tub and do not wish to rebuild any of the walls, you must get an exact replacement tub.

REMOVING THE OLD TUB

Most, if not all old tubs are made of cast iron. You can fairly easily break them up with a sledgehammer. However, it is best to remove the pipes and the tile and Sheetrock atop the tub's lip before you start banging away. Do it slowly and carefully, section by section.

With the old tub out of the way, clean up the contacting walls and floor so that the new tub can rest directly against the studs and directly atop the subfloor.

INSTALLING A NEW TUB

There are three types of tubs manufactured today: cast-iron, steel and plastic. Both the cast-iron and the steel have a porcelainized surface. In appearance they are identical. The only difference discernible between them is produced by banging. The steel has a hollow sound. The cast-iron makes no sound at all. The cast-iron tub is very heavy and expensive. The steel tub is far lighter, is less costly and will last 25 or more years. The plastic tub is the least expensive, lightest in weight and usually loses its original shine with time.

No matter which tub you secure, protect it with multiple layers of wet newspaper, which you can hold in place with flour-and-water paste. And don't drag it across the floor—the bottom is easily chipped.

Connecting an exact replacement—With the old tub out of the way, with the studs and floor cleaned up and with the drain and other pipes disconnected, a couple of boards are laid on the floor and the new tub is carefully slid into place.

With the tub in place, the missing wall Sheetrock, wall tile and floor tile are replaced. Plumber's putty is used to seal the drainpipe connection to the tub. The drain and the balance of the piping is connected and the job is done. The weight of the tub and the tile surrounding it keeps it in place.

Connecting a modern tub—Modern tubs have two holes, one for the drain and the other for the pop-up valve control. The water spout and its valves are always mounted above the tub. When an old-style tub is replaced with one of modern design, the pocket or space the new tub is to occupy must have clearance for the pop-up valve tube and connections. The spout, valves and shower-head supports and stub-outs (short pipes that project through the finished wall and carry the spout and shower head) must be positioned and connected before the tub is moved in. Some of the drain piping can be done prior to positioning the tub, but the final drain-to-tub connections usually have to be made with the tub in place, so leave plenty of working space.

Installing a new valve assembly—The valves or controls for a tub and shower consist of a lumpy brass casting with valves and water connections at the ends, a diverter valve in the middle and sweat (solder) connections top and bottom for the shower and spout respectively. Turning the central diverter valve directs the flow of water upward to the shower or downward to the spout.

The assembly is fastened with brass straps and copper nails to a wooden cross brace rabbeted firmly into the wall studs. A pipe runs upward from the center of the assembly to the shower-head connection, which is an elbow also strapped to a cross brace. A second and short pipe runs downward from the center of the assembly to a second elbow which accepts the stub-out (pipe) onto which the spout screws. The spout elbow is also strapped down to a cross brace. Brass straps and copper nails are used to eliminate the possibility of rapid galvanic corrosion that would occur if iron straps and nails were used against copper pipe and both became wet.

Whether threaded joints or sweated joints are used to make the aforementioned connections depends on faucet assembly casting. If it has been made to be soldered to its supply pipes, remove the valve stems before soldering.

Generally the spout, valves and shower head

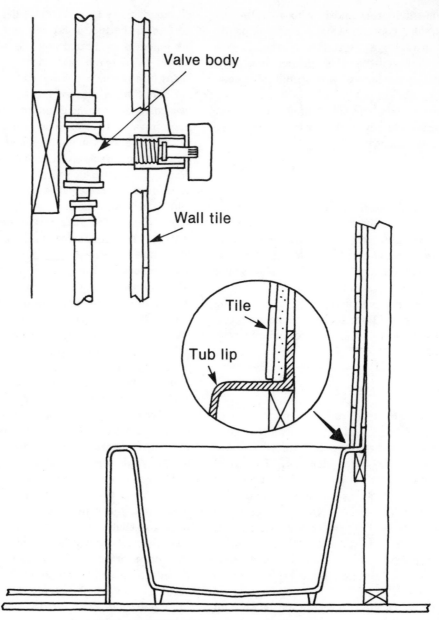

178. Tub installation and connection details.

are mounted so that they emerge from between tiles. This reduces the amount of tile cutting necessary. To do this, just remember that wall tile is usually 4-inch squares and mount your casting and elbows accordingly.

The position of the valves and the spout in relation to the tub is not at all critical. So long as the spout and the shower head point into the tub and the handles are convenient, everything will work fine. The distance from the front surface of the cross brace to the front surface of the in-place wall tile must be such that faucet knobs and escutcheons fit properly. If the casting is too far toward the tub, there will be free space beneath the handles. If the casting is too far rearward, the knobs may not fit.

The way to secure this dimension is to temporarily mount the casting on a board, install the escutcheon and knobs and then measure the distance from the underside of the escutcheon to the board. If, for example, the distance is 4 inches and you are using ceramic tile on ½-inch Sheetrock, you know that the front surface of the brace must be 4 inches less ⅜ inch for the tile and ½ inch for the Sheetrock, or 3⅛ inches back from the front side of the studs (underside of the Sheetrock).

Another minor dimensional problem is that of the spout-supporting pipe (stub-out). This nipple (short length of threaded pipe) must be positioned so that the spout tightens with its nozzle pointing downward. By assembling the spout and nipple and measuring from the rear of the spout to the rear of the elbow, you can determine just how the nipple supporting the spout must be set.

179. The usual method employed to mount valves and shower-head assembly in a wall.

CONNECTING THE WATER PIPES AND DRAINPIPES

The water pipes can be connected before the tub is slid into place. A portion of the drain can sometimes be assembled prior to installing the tub, but the last few connections usually can only be done with the tub in place.

If the original pipes are threaded, unscrew them and replace them with nipples followed by adapters. Use tubing to go from the adapters to the valve assembly. If the assembly is made for threaded pipe, use adapters to go from the tube to the assembly. This will cost more than going from the old threaded pipe to a threaded assembly, but in most instances it will be lots easier. Use ½-inch tubing to supply the tub and shower.

Before sliding the tub in place and assembling the drain, make certain that you have sufficient room to work. Don't hesitate to enlarge the hole cut for the drainpipe. That portion of the flooring supports nothing.

Always install a trap in the tub's drain and always leave some means of access to the trap. Sooner or later work will have to be done here. At the very least, mark the ceiling or wall where you may have to cut through at some future date.

Position the tub, then use a little soft putty above and below the metal lip and the compression nut holding the drainpipe to the tub's bottom. Make the nut snug, but not tight. Overtightening can crack the porcelain.

The shower head screws onto the brass nipple stub-out projecting into the room from a second cross brace.

INSTALLING A TUB OF A DIFFERENT SIZE

So far we have discussed the steps involved in removing an old tub and replacing it with a more or less physically identical tub. Comparatively little was done to the walls. The studs (wall frame) were not touched.

When the size or shape of the new tub is different from the old, one or more walls have to be rebuilt. When all you need to do is to bring the old wall out an inch or so it is usually easier to use furring strips. When the old wall has to be moved back, it is always easier to take the old wall down and rebuild it where needed than attempt to "hack off" an inch or so.

Constructing the new walls—To secure a more accurate "fit" between the new tub and the new walls, use the tub itself as a guide. Cover the new tub with many layers of paper pasted into place. Install the pop-up valve mechanism and then carefully slide it into the exact spot it is to occupy. Use a level to make certain that the tub is level or pitches slightly toward its drainpipe. Use the level to position the sole plate (which carries the studs) alongside the tub. Pencil the sole plate's position on the floor. Remove the tub. Nail the sole plate in place. Install the studs and top plate. Follow with the braces for the faucets, spout and shower head. Install the plumbing. Move the tub back into place and connect the drainpipe.

Bear in mind that a steel tub or a plastic one has to be supported for most of its length, not just at the corners, and that the Sheetrock overlaps the lip of the tub.

Installing tile—There is only one precaution. Just leave enough space around the valves to permit the entrance of a socket wrench for service.

180. **Primary steps in installing a tub**

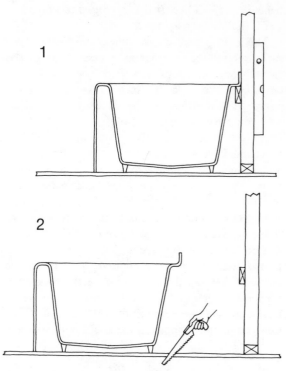

1. Position the tub on the subflooring.
2. Move the tub out of harm's way. Cut a hole in the floor for the pipes.

CHAPTER THIRTEEN

Installing Appliances

A dishwasher requires a supply of hot water and a drain connection. A clothes washer needs both hot- and cold-water piping as well as a drain. A food disposal unit or grinder requires no plumbing work, though grinders usually have provisions for connecting a dishwasher drainpipe.

Hot- and cold-water supplies present no problems. You can run your lines using galvanized, copper or plastic, as you see fit. The drain, however, can be a major plumbing job if you have no nearby sink or cellar-floor drain to which you can connect your appliance.

In the latter case you have the option of installing a dry well or, if you are installing a clothes washer, connecting to an existing drainpipe. A dishwasher can only be connected to a drainpipe which leads to the sewer. It cannot be emptied into a dry well because the food particles will rot and produce an odor.

Traps and vents—As stated previously, a trap prevents sewer gas from entering the building through the drain opening in the appliance or fixture. When you connect your appliance to a fixture or a cellar-floor drain, the trap provides a protective plug of water. There is no chance of sewer gas and odor entering your home.

When you connect a fixture or an appliance directly to an existing drainpipe you must always interpose a trap, and under certain conditions you must vent that trap. The trap prevents sewer gas from entering and the vent prevents the protective plug of water from being drained by a high wind or the simultaneous discharge of several fixtures into the common drainpipe.

Exactly how traps and vents are to be connected to provide maximum protection is spelled out in the plumbing code. By necessity, the following instructions cover only a portion of the code and give only a limited example of its application. Your local plumbing inspector will help you with any questions you may have concerning your particular plumbing installation.

WATER SUPPLY

Although a shutoff valve in the water supply pipeline(s) increases material cost and labor, it is a necessary investment. There will come a time when it will be necessary to repair or remove the machine. If you cannot shut off the appliance water line(s) you will be forced to shut down all or a portion of the house's water supply for the length of time the device is out of service. And in an emergency, which is not uncommon with dishwashers, a shutoff valve can save you a minor flood.

Connecting to galvanized—Unless there is a union in the line nearby, you have to cut the pipe with a hacksaw or a pipe cutter. The halves of the original pipe are then unscrewed and replaced with a nipple, a union, a nipple, a Tee and a third nipple. If you can have the original pieces of pipe cut and threaded to size you can save a few dollars. Directions on how to work with galvanized pipe are given in Chapter Four.

Connecting to copper tubing—If the tubing is rigid, use a hacksaw or a pipe cutter to remove a section of tubing several inches longer than the

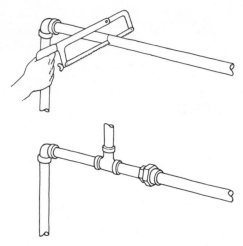

181. To tap on to a water line of galvanized pipe, cut the pipe at an angle with a hacksaw, then cut the old pipe into nipples of the proper length or purchase the necessary nipples plus a Tee to make the assembly shown.

182. The first step in tapping into the middle of a run of plastic pipe. Lay your Tee alongside and mark off the section of pipe that must be cut out to admit the Tee.

Tee to be connected. Sweat the Tee to one end of the original tube. Use the cut section of tubing and a slip fitting to join the other side of the Tee to the end of the other portion of the line. To do so you have to shorten the piece of tubing you removed. Cut it so that it is ¼-inch short of contact when you slip it into the Tee and line it up with the end of the water line.

If you are going to tap into soft tubing and have several feet of tubing to either side of the point at which you want to insert the Tee, cut the tubing at that point. Bend a shallow offset into one end of the cut tube. Sweat a Tee onto that end. Now bend another shallow offset into the remaining pipe end and sweat that to the Tee.

Directions on how to work with copper tubing are given in Chapter Five.

Connecting to plastic pipe—The same technique that is used with rigid copper is used with plastic. A section of tubing is cut out of the water line and it is replaced with a Tee, a plastic slip coupling and the original section of plastic pipe, which is shortened. The slip fitting is slid up over the pipe end. The short section of pipe is inserted and solvent-welded to the Tee. Then the slip fitting is slid down over the remaining joint and solvent-welded in place. Directions on how to work with plastic pipe are given in Chapter Six.

183. Cut the plastic pipe on your marks.

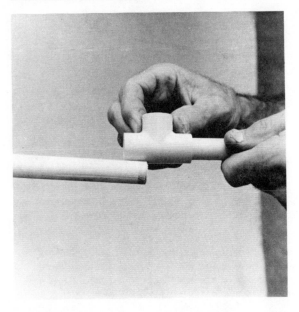

184. Cement your Tee to one end of the cut pipe.

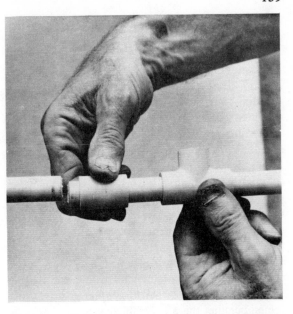

186. Connect the recently cut short piece of pipe to the Tee. Pass a slip fitting over the end of the newly cemented-on piece.

185. Cut a few inches off the end of the other pipe. Save the piece.

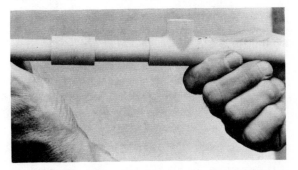

187. Apply cement to the ends of the pipe and slide the slip fitting over the two ends.

Extending the Tee(s)—The line from the Tee can be any practical distance necessary. However, if you have to run over a couple of feet it is inadvisable to run anything smaller than ¾-inch galvanized or ½-inch copper or plastic.

VALVES AND CONNECTIONS

Washing machines usually come equipped with rubber hoses terminating in female water-hose couplings. A **water-hose coupling** is a coarse-threaded coupling that screws onto faucets having water-threads cut (or molded) into their spouts. Thus, you can use a water-hose faucet or a **sill cock** (same thing, but shaped a little differently) as both a means of connecting your machine to the water lines and a means of shutting off the water when necessary. This is the connection and control method almost always used with clothes washers.

Suitable faucets with screw threads and sweat connections are available. To use these faucets with plastic pipe a threaded adapter is used. Just remember that there is a rubber washer that goes inside the hose coupling. If it isn't there the joint will leak very badly.

Dishwashers present a slightly different problem because their single water-feed pipe comes equipped with a variety of terminations. For some you will need a special fitting to go from the rubber hose to threaded pipe. In some instances you will find that the easy way out consists of cutting the end off the hose and clamping it onto your feed pipe.

In any case, a standard ½-inch shutoff valve should be installed in the line extending from your Tee. To use the clamp arrangement, connect a threaded, ⅜-inch nipple to the hot-water supply. Force the hose end over the nipple and use a strap clamp to hold it in place.

CONNECTING THE DRAINPIPE

Dishwashers are usually installed in the kitchen near the sink. In such cases it is relatively easy to connect the appliance to the sink's drain. This is done by making the connection ahead of the sink's trap. In that way the single trap protects both the sink and the washer.

The thin-wall brass pipe connecting the trap to the sink is removed and replaced with a washing machine Tee. You may have to cut the Tee a bit to make it fit. Use a fine-toothed hacksaw or a pair of compound tin snips.

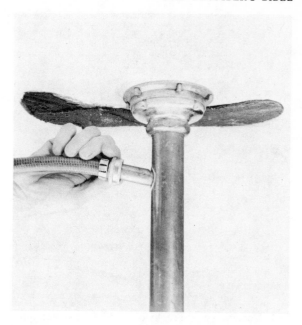

188. Installed washing-machine Tee. Hose is connected to coarse thread on smaller-diameter pipe, as shown.

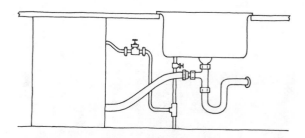

189. How a dishwasher may be connected beneath a counter top.

The washer's drainpipe connects to the side arm on the Tee. It has coarse threads. If your washer's drainpipe has the proper fitting, just screw it on. If not, secure the proper fitting or force the hose over the threads and use a strap clamp to hold it in place.

When the aforementioned is not practical and the dishwasher or the clothes washer is above a basement washtub, the appliance can easily be drained directly into the tub. A hole is drilled through the floor. A length of ¾-inch or larger garden hose is slipped down through the hole to the basement beneath. The lower end of the hose is placed inside the tub. The upper end of the hose is connected to the appliance drainpipe by means of suitable fittings or a short section of copper tubing and two hose clamps. The appliance drains into the tub. The air separation between the end of the hose and the tub's drainpipe is called an **air break.** This provides the necessary air vent in the drain connection.

Using a floor drain—When there is no tub conveniently beneath the appliance to be drained, but there is a floor drain, the appliance can be drained into the floor drain. All properly installed floor drains have built-in traps. Thus, by using the air-break principle you can utilize a floor drain for appliance drainage.

A clearance hole is cut into the cast-iron drain cover. The garden hose is laid on the cellar floor. To keep it from being kicked away it can be strapped down with rawl plugs and brass screws. You will need a star drill to cut the holes in the concrete. After that, rawl plugs are pushed into the holes. The plugs hold the screws.

Using a dry well—A dry well consists of a hole in the ground filled with loose stones and covered with soil and grass. Water to be disposed of is led into the well, where the water has the time necessary to seep into the earth. Dry wells have long been used to rid homes of rainwater collected by a roof. They work just as well with clothes-washer water as with rain.

Building a dry well—Dry wells work best in sandy soil; they work poorly when the soil is heavy with clay and they don't work at all when the soil is continuously waterlogged.

For the average household a well about 5 feet in circumference and 7 feet deep should be more than sufficient; the top two feet should be filled with earth. Use any loose, rough stone you can find. The purpose of the stone is to support the roof without filling the hole up. Use the largest stones on the bottom. Use flat stones to bridge round stone and use the smaller stones on top. Install your drainpipe end about midpoint in the well. Cap the well with a layer of tar paper and two feet of earth.

Use 1-inch or larger flexible plastic pipe from the well to the appliance. Run the pipe below the

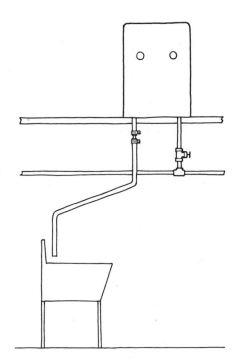

190. How a sink or a tub on a lower level may be used for drainage.

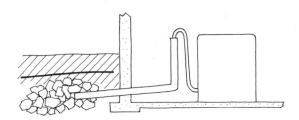

191. How a dry well may be constructed to serve as a drain for a clothes washing machine.

frost line and bring it inside the building through a hole in the floor so that all of the pipe outside the building is protected from the frost by a layer of soil. Use an adapter and terminate the inside end of the 1-inch pipe in a vertical section of 1½ or 2-inch pipe. Do this so that you can hook your appliance drain over the edge of the plastic pipe and have a little space for air around it.

From this point on out there is nothing to do until the well fills up with lint. By that time, something better than a washing machine may have been invented.

Remember that a dry well never should be used with a dishwasher. The collected particles of food will rot and cause gas and odors.

VENTING REQUIREMENTS

As stated previously, all fixture and appliance drains must be protected from sewer gas by a trap. All traps must be vented. However, all traps need not have an individual vent pipe. Under certain conditions traps can be vented by the soil stack, which is the large vertical pipe into which the fixtures drain.

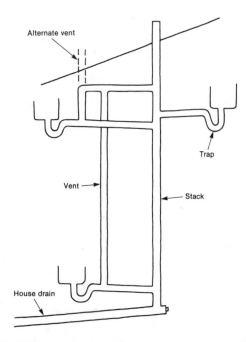

192. Basic code requirements: A top-floor fixture may be connected to the stack without venting if the trap is no more than about 5 feet from the stack and drainpipe pitch is about ½ inch to the foot. If the top-floor fixture's trap is farther than 5 feet from the stack, fixture drain must be vented at 5 or less feet. Vent may connect to stack 1 or more feet above fixture, or may go through roof. On all floors below the top, each fixture or appliance trap must be vented, either to the stack at a point 1 or more feet above highest fixture or to an individual or combined (with other vents) pipe passing through the roof.

Stack venting—When an appliance or fixture on the top floor (where there are no higher appliances or fixtures) is connected a certain way, the stack itself acts as a vent.

Specifically, when 1¼-inch pipe (minimum) is used to drain a dishwasher and the distance from the trap **weir** (water level on the outside bend) is no more than 5 feet to the stack and the pipe's pitch is no more than ¼ inch to the foot, no vent pipe is required. When 1½-inch pipe (minimum) is used for a clothes washer and the pitch is no more than ¼ inch to the foot, no vent pipe is needed if the distance is 6 feet or under.

Phrased differently, pitch should be kept between ¼ inch and ½ inch per foot and the point where the drainpipe enters the stack should be no more than equal to the pipe's diameter.

When the pitch is greater, or when the length of the pipe is greater, so that the end of the pipe where it connects to the stack is lower than the end connected to the appliance, the trap will siphon (empty) itself. There will be no protective plug of water inside.

If and when you must have a longer run of drainpipe to the stack, the drainpipe must be vented 5 feet or less from the trap.

Connecting the vent pipe—The vent pipe should not be less than 1 inch for the two pipe sizes mentioned. For larger sizes the vent should be at least half the drainpipe size, but never less than 1 inch.

The top of the drainpipe must either pass through the roof for a distance of at least 1 foot, or it must be connected to another, higher vent pipe or it must be connected to the soil stack at a point at least 1 foot above the top of the highest fixture connected to the soil stack. The vent pipe can twist and turn, but it must always angle upward and never down. The vent pipe can be of any piping material; it must be joined as tightly and as permanently as pipe under pressure and must be increased to 3 or more inches in diameter when it passes through the roof. This is necessary to keep the top of the vent from closing up with frost in the winter.

Connecting to a cleanout—The easy way is to connect the new drain to a cleanout in the existing drainpipe. You will find cleanout plugs wherever a drainpipe makes a sharp turn and every 45

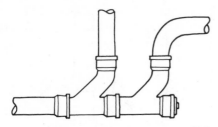

193. How a second cleanout may be added to an in-place cleanout to permit the connection of a second soil pipe or soil stack (soil pipes are connected to toilets; waste pipes drain sinks).

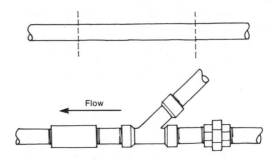

194. How a new waste pipe may be connected to an existing waste pipe. The old pipe is cut and threaded. A drain Wye is then connected in the line with the aid of nipples and a union. Note the relation of the Wye to the direction of waste-water flow.

feet along the length of the drain. This is what the code calls for, and if your home was "plumbed" to the code, you will find these cleanout plugs.

The plug is removed and a threaded cleanout fitting is connected in its place. The added fitting enables you to connect your new drainpipe. The plug on the new fitting enables you to clean the drainpipe should the need arise.

The only trouble with using an existing cleanout is that they are not usually where you need them. Be advised, however, that it is far easier to run extra drainpipe to utilize a cleanout than it is to cut into a drainpipe.

Connecting to a branch drain—A branch drainpipe connects one or more fixtures on a floor with the soil stack (the vertical drainpipe). So long as the branch drain is 1½ or more inches in size it will accommodate your appliance without any difficulty.

If the branch is galvanized, you have to cut it, remove the pieces and install a union and a drain Wye. Use nipples or have the old pieces of pipe cut and threaded to proper length. See Chapter Four for directions on working with galvanized pipe. The Wye must be a drain Wye or it will catch and hold debris, and it must point toward the sewer (downward); otherwise the waste or soil will enter the fitting and will head up toward your appliance.

If the branch drain is copper DWV pipe, you have to cut the pipe and remove a section. The section is replaced with a copper drain Wye, a short section of copper drainpipe and a slip coupling, which are sweated in place.

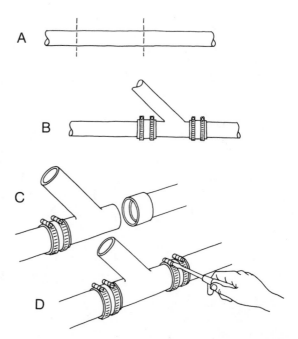

195. How No-hub pipe joints may be used to add pipe connections to any existing thick-wall pipeline, as in the change from A to B above. This includes galvanized, brass and cast-iron pipelines. First, a section of the in-place pipe, 1 inch longer than the fitting to be added, is cut out of the pipe. The fitting, in this case a drain Wye, is positioned and held in place with a neoprene sleeve and a No-hub clamp (C). Finally, the second No-hub clamp is tightened in place (D).

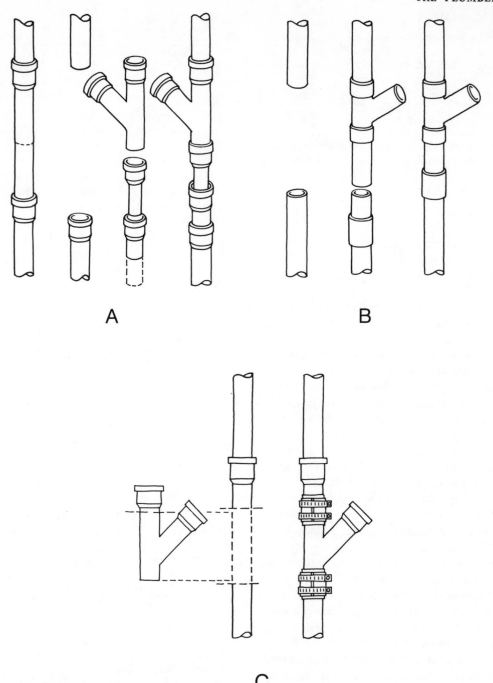

A

B

C

196. A How a Sisson fitting is introduced into an existing soil stack. This fitting is
     used where local codes will not permit the use of No-hub fittings.
   B How a plastic or copper drain Wye can be added to an existing drain stack or
     to a waste stack or pipeline.
   C How a standard, cast-iron drain Wye can be modified and installed within the
     length of an existing soil stack with the aid of two No-hub fittings.

If the branch drain is plastic pipe, the same procedure is used, but the parts are solvent-welded in place.

Chapters Five and Six, respectively, discuss working with copper and plastic.

Connecting to a soil stack—This can be difficult, especially if the stack is of cast iron. When you connect to a branch drain you are joining your appliance drain to a horizontal pipe, which you can cut without worrying about its support. At worst you can fasten some extra straps to make certain that it doesn't come down. When you cut into a vertical stack it is difficult to be certain that the stack is suitably supported, and it is difficult to support the stack, should it be necessary—especially if you are dealing with a cast-iron stack that may weigh a ton or more.

Since a connection is most easily made to plastic drainpipe, we will start with that. The pipe immediately above the area to be cut is cleared. If there is no pipe clamp, one is attached. Connect it beneath a fitting or a hub. Next cut a section out of the stack, add a drain Wye, a section of stack pipe and a slip coupling. Solvent-weld them in place. Connect your drainpipe and you are done.

To connect to a copper drain stack the same procedure is followed, but of course the pipe and fittings are sweated in place.

To connect to cast iron you must make doubly certain that a strong support is present. If you aren't certain or if there is no support, make one of a pair of 2×4's stood on end, with a pair of cross braces nailed to either side of a hub, and then wired to the hub with several turns of galvanized wire. Nail the top and bottom of the 2×4's in place.

You will have to hacksaw a section out of the cast-iron pipe, which isn't difficult, but takes time. The size of the section you remove will depend on the Wye you insert. The easiest is a No-hub Wye. To insert this type of fitting you remove a section about ½-inch larger than the fitting. The fitting is lined up with the pipe and held in place with neoprene sleeves and stainless-steel clamps. The connection from the Wye to your new drainpipe can be made with a similar fitting or any other fitting that is convenient.

The trouble with a No-hub fitting is political.

All local plumbing departments will not permit this easy means of joining cast-iron pipe. In such cases you have to go the difficult route, and use a Sisson fitting. This consists of two sections of cast-iron pipe made so that one section can slip in and out of the other to vary the overall length. To install a Sisson fitting you have to remove a section of pipe large enough to insert the cast-iron drain Wye plus the two-section Sisson fitting. The cut must be made so as to include the bottom section (spigot) of the drainpipe. Thus when the section is out you have a hub on the bottom and a straight section (spigot) on top. The Wye and the Sisson are assembled and slipped into line with the rest of the pipe; then the Sisson is expanded. Now you must lead-caulk four joints.

Start by filling the pocket (the space between the pipe end and the hub) with oakum. Wrap it several times around the pipe and drive it home with a caulking iron, which is a chisel with a very blunt (¼-inch thick) point. Drive sufficient oakum into the pocket to fill it within ¾ inch or so of the top of the hub. Follow with a layer of lead wool, which is lead cut into fine strips much like tinsel. Braid a rope with the tinsel and pound it into place just as you did with the oakum. When you have a lead mass flush with the top of the hub, the job is done. The oakum seals the joint; the expanded lead holds the oakum in place.

The joint between the cast-iron Wye and your drainpipe will depend on the type of drainpipe you are running. If it is threaded galvanized pipe, you can secure a Wye with a threaded opening and make the joint with the aid of a nipple and a union. With plastic pipe you can make the joint between the plastic and the cast iron using a special compound made for the purpose. The plastic is inserted into the fitting and the joint is sealed with the compound. When you are running copper, you have to use a transition joint between the cast iron and the copper. This joint is lead-caulked into the cast iron and then sweated to the copper drain.

CONNECTING A FOOD DISPOSAL UNIT

Food disposal units are installed between the kitchen sink drain and the drainpipe. The tail pipe, the strainer and the connection are removed and replaced with a special strainer supplied with the disposal. There are various designs, but the method of connection becomes obvious as soon as you examine the parts.

The connection between the disposal outlet and the sink drain is usually most easily made with a section of rubber hose and a pair of clamps.

The small pipe up near the top of the unit is designed for a dishwasher drain connection. If you have a dishwasher, the rubber dishwasher hose is forced over the open pipe end and clamped in place. If you do not have a dishwasher to connect, this opening must be sealed. Clamp a length of hose over the opening, fold down the other end of the hose and wire it snugly against the pipe end to seal the opening.